Series Editor

BEYOND THE COMMON CORE

A HANDBOOK FOR

Mathematics

in a PLC at Work™

LEADER'S GUIDE

Solution Tree | Press
a division of
Solution Tree

Timothy D. Kanold
Matthew R. Larson

A Joint Publication With

**NATIONAL COUNCIL OF
TEACHERS OF MATHEMATICS**

555 North Morton Street
Bloomington, IN 47404
800.733.6786 (toll free) / 812.336.7700
FAX: 812.336.7790

email: info@solution-tree.com
solution-tree.com

Visit **go.solution-tree.com/mathematicsatwork** to download the reproducibles in this book.

Printed in the United States of America

19 18 17 16 15   1 2 3 4 5

Library of Congress Cataloging-in-Publication Data

Kanold, Timothy D.
   Beyond the common core : a handbook for mathematics in a PLC at work. Leader's guide / authors, Timothy D. Kanold and Matthew R. Larson ; editor, Timothy D. Kanold.
      pages cm
   Includes bibliographical references and index.
   ISBN 978-1-936763-62-7 (perfect bound) 1. Mathematics--Study and teaching (Continuing education)--United States. 2. Curriculum planning. I. Larson, Matthew R. II. Title.
   QA13.K35 2015
   510.71'273--dc23
                                    2014049301

Solution Tree
Jeffrey C. Jones, CEO
Edmund M. Ackerman, President

Solution Tree Press
President: Douglas M. Rife
Associate Acquisitions Editor: Kari Gillesse
Editorial Director: Lesley Bolton
Managing Production Editor: Caroline Weiss
Senior Production Editor: Suzanne Kraszewski
Copy Editor: Sarah Payne-Mills
Proofreader: Elisabeth Abrams
Text and Cover Designer: Laura Kagemann
Text Compositor: Rachel Smith

This leader's guide is dedicated to the greatest school leader and practitioner I have ever known: Dr. Richard DuFour. I have had the privilege to work with Rick for more than thirty-five years. He has impacted my life at the deepest level, just as he has for thousands of others who have been part of the Professional Learning Communities at Work™ movement—a movement he created, nurtured, taught, and led not just at Adlai E. Stevenson High School District 125 in Lincolnshire, Illinois, but across the entire globe. His genius and passion as a writer, teacher, inspirational leader, and coherence builder are without compare in K–12 education and are true gifts to us all. It would be our gift back to him if we could each find the courage to be as passionate about our work, as committed to the cause of student learning, and as dedicated to working collaboratively to build great schools as Rick, while also leading with a humble spirit and heart. May his many lessons serve each of us well in life's journey as teachers, teacher leaders, and creators of the school cultures necessary to achieve Rick's daily challenge to "soar and not settle."

Acknowledgments

Once again, I find myself thanking Tim Kanold for his vision, positive perseverance, and sincere commitment to improve mathematics education in the United States. Your leadership, mentorship, and friendship are deeply appreciated. A sincere thank you to the leaders and teachers of mathematics in the Lincoln Public Schools, particularly Delise Andrews, Susie Katt, Julie Kreizel, and Jerel Welker, who, over the last two decades, have provided critical feedback on the recommendations in this handbook and challenged me as we collaborated to improve student learning. Finally, I owe a special thanks to Tammy, my wife, who always supports and encourages my professional endeavors, often at her own personal sacrifice.

—Matthew R. Larson

First and foremost, I extend my thanks to Matt Larson for his incredible leadership, humor, and friendship in my life. Matt understands the joy, the pain, and the hard work of the writing journey, and he is one of the most talented writers I have had the privilege to know. My thanks to our expert reviewers, who have dedicated their lives to the work and effort described within the pages of this handbook and whose feedback has influenced almost every page. Thanks, too, to Jeff Jones and Douglas Rife from Solution Tree for their belief in our vision and work in mathematics education, and of course to Suzanne Kraszewski, editor extraordinaire, who made this entire series become a reality. Special thanks to Rick DuFour for his wisdom, insight, advice, and deep understanding of what it takes to become an authentic professional learning community. My thanks also to my wife, Susan, my loving critic, who understands how to formatively guide me through a handbook series project as bold as this.

—Timothy D. Kanold

Solution Tree Press would like to thank the following reviewers.

Bill Barnes
Coordinator of Secondary Mathematics
Howard County Public Schools
Ellicott City, Maryland

Diana Kasbaum
Mathematics Education Consultant
Association of State Supervisors of Mathematics, Immediate Past President
Madison, Wisconsin

Janice L. Krouse
Mathematics Faculty/Curriculum and Instruction Consultant
Illinois Mathematics and Science Academy
Aurora, Illinois

Donna Simpson Leak
Superintendent
Community Consolidated Schools District 168
Sauk Village, Illinois

Suzanne Mitchell
Mathematics Consultant
Past President, National Council of Supervisors of Mathematics
Jacksonville, Arkansas

Ed Nolan
Director, Mathematics Implementation and Development Team
Montgomery County Public Schools
Rockville, Maryland

Kit Norris
Senior Associate
Mathematics at Work™ Learning Group
Boston, Massachusetts

Sue Pippen
Pippen Consulting
Past President, Illinois Council of Teachers of Mathematics
Plainfield, Illinois

Connie Schrock
Professor of Mathematics and Economics
Emporia State University
Emporia, Kansas

Sarah Schuhl
Senior Associate Mathematics at Work™
PLC Associate and Mathematics Educational Consultant
Portland, Oregon

Nanci Smith
Mathematics Consultant and Senior Associate Mathematics at Work™
Effective Classrooms Educational Consulting
Cave Creek, Arizona

John Staley
Director of Mathematics, PreK–12
Baltimore County Public Schools
Baltimore, Maryland

Becky Walker
Director of Curriculum
Appleton Area School District
Appleton, Wisconsin

Gwen Zimmermann
Assistant Principal of Teaching and Learning
Adlai E. Stevenson High School
Lincolnshire, Illinois

Visit **go.solution-tree.com/mathematicsatwork** to download the reproducibles in this book.

Table of Contents

About the Series Editor

Timothy D. Kanold, PhD, is an award-winning educator, author, and consultant. He is former director of mathematics and science and served as superintendent of Adlai E. Stevenson High School District 125, a model professional learning community district in Lincolnshire, Illinois. He serves as an adjunct faculty member for the graduate school at Loyola University Chicago.

Dr. Kanold is committed to a vision for Mathematics at Work™, a process of learning and working together that builds knowledge sharing, equity, and excellence for all students, faculty, and school administrators. He conducts highly motivational professional development leadership seminars worldwide with a focus on turning school vision into realized action that creates increased learning opportunities for students through the effective delivery of professional learning communities for faculty and administrators.

He is a past president of the National Council of Supervisors of Mathematics and coauthor of several best-selling mathematics textbooks. He has served on writing commissions for the National Council of Teachers of Mathematics and the National Council of Supervisors of Mathematics. He has authored numerous articles and chapters on school mathematics, leadership, and professional development for education publications.

In 2010, Dr. Kanold received the prestigious international Damen Award for outstanding contributions to the leadership field of education from Loyola University Chicago. He also received the Outstanding Administrator Award from the Illinois State Board of Education in 1994 and the Presidential Award for Excellence in Mathematics and Science Teaching in 1986.

Dr. Kanold earned a bachelor's degree in education and a master's degree in applied mathematics from Illinois State University. He completed a master's in educational administration at the University of Illinois and received a doctorate in educational leadership and counseling psychology from Loyola University Chicago.

To learn more about Dr. Kanold's work, visit his blog *Turning Vision Into Action* at http://tkanold .blogspot.com, or follow @tkanold on Twitter.

To book Dr. Kanold for professional development, contact pd@solution-tree.com.

About the Authors

Matthew R. Larson, PhD, an award-winning educator and author, is the K–12 mathematics curriculum specialist for Lincoln Public Schools in Nebraska and president-elect of the National Council of Teachers of Mathematics. Dr. Larson has taught mathematics at the elementary through college levels and has held an honorary appointment as a visiting associate professor of mathematics education at Teachers College, Columbia University.

He is coauthor of several mathematics textbooks, professional books, and articles in mathematics education. A frequent keynote speaker at national meetings, Dr. Larson's humorous presentations are well-known for their application of research findings to practice.

Dr. Larson earned a bachelor's degree and doctorate from the University of Nebraska–Lincoln.

To book Matthew R. Larson for professional development, contact pd@solution-tree.com.

In addition to being the series editor, **Timothy D. Kanold, PhD,** is a coauthor on this book.

Introduction

You have high impact on the front lines as you snag children in the river of life.

—Tracy Kidder

Your work as a superintendent, principal, assistant principal, school site administrator, central office mathematics program leader, mathematics instructional coach, grade-level or course-based team leader, or central office administrator and leader is one of the most important and, at the same time, most difficult jobs to do well in education. Since the release of our 2012 Solution Tree Press series, *Common Core Mathematics in a PLC at Work*™, our authors, reviewers, school leaders, and consultants from the Mathematics at Work™ team have had the opportunity to work with thousands of school leaders, leadership teams, and teacher teams across the United States who are just like you: educators trying to urgently and consistently seek deeper and more meaningful solutions to a sustained effort for meeting the challenge of improved student learning in mathematics.

From California to Virginia, Utah to Florida, Oregon to New York, Wisconsin to Texas, and beyond, we have discovered a thirst for implementation of K–12 mathematics programs that will sustain student success over time. Your leadership serves a significant contribution to the K–12 solution and effort. This handbook describes how, just as each teacher must teach and lead his or her students, there is a parallel expectation that you will teach and lead your faculty. The teacher might be on the front line, but you are the broker of the support, accountability, resources, inspiration, time, and energy necessary for teacher success. They need you.

Certainly the Common Core State Standards (CCSS) have served as a catalyst for much of the national focus and conversation about improving student learning. However, your essential work as a school leader in your local school and district takes you *beyond* your states' standards—whatever they may be. As the authors of the National Council of Teachers of Mathematics ([NCTM], 2014) publication *Principles to Actions: Ensuring Mathematical Success for All* argue, standards in and of themselves do not describe the essential conditions necessary to ensure mathematics learning for all students. You, as a leader of the mathematics teaching faculty and staff in your school or district, serve the critical role of ensuring the essential conditions for success are in place.

Thus, this mathematics teaching and assessing handbook is designed to take you *beyond the product* of standards themselves by providing you with the guidance, support, and leadership tools necessary to help the adults in your system achieve mathematics program greatness within the context of higher levels of demonstrated student learning and performance.

Whether your leadership work occurs in a state that is participating in one of the CCSS assessment consortia or in a state that uses a unique mathematics assessment designed only for your state, it is our hope this handbook provides a continual process that allows your mathematics program to become one of great mathematics teaching and learning.

Your daily leadership work begins by understanding there are thousands of instructional and assessment decisions your teachers and teacher teams (those adults closest to the action) will make every day in every unit. Do those decisions make a significant difference in terms of increased levels of student achievement? Your leadership role is to make sure they do.

The Grain Size of Change Is the Teacher Team

We believe the best strategy to achieve the expectations of the CCSS-type state standards is to create schools and districts that operate as professional learning communities (PLCs), and, more specifically, within a PLC at Work™ culture as outlined by Richard DuFour, Rebecca DuFour, Robert Eaker, and Tom Many (2010). We believe that the PLC process and culture support a grain size of change that is just right—not too small (the individual teacher) and not too big (the district office)—for impacting deep change. The adult knowledge capacity development and growth necessary to deliver on the promise of your state standards reside in the engine that drives the PLC school culture: the teacher team. In your sphere of influence, your primary role is to monitor, provide feedback, and expect team action.

There is a never-ending aspect to your professional leadership journey and your ability to help teacher teams know how to become more transparent in their practice. As John Hattie (2012) states in *Visible Learning for Teachers: Maximizing Impact on Learning*:

> My role as a teacher is to evaluate the effect I have on my students. It is to "know thy impact," it is to understand this impact, and it is to act on this knowing and understanding. This requires that teachers gather defensible and dependable evidence from many sources, and hold collaborative discussions with colleagues and students about this evidence, thus making the effect of their teaching visible to themselves and to others. (p. 19)

Thus, your role as a school leader and teacher is to also "know thy impact" on teacher/adult learning in your school. Your leadership role (Warren Little, 2006) is to create the conditions necessary for adult knowledge capacity building through a professional development experience that includes four main characteristics. The teacher professional development experience should be:

1. Ongoing and sustained rather than episodic

2. Collective rather than individualistic

3. Job-embedded so learning is at point of use

4. Results-oriented with activities linked to improved student achievement

Your work then, as a school leader, is to push hard to create an open and transparent school culture, monitor the effectiveness of teacher team discussions and actions, and focus the teacher teams' work on results. This handbook provides you with the tools to create such a culture. Our experience reveals this work will help you in other areas of curriculum leadership as well.

Knowing Your Vision for Mathematics Instruction and Assessment

Quick—you have thirty seconds: turn to a colleague and declare your vision for mathematics instruction and assessment in your mathematics department and in your school. What exactly will you say?

More importantly, on a scale of 1 (low) to 6 (high), what would be the degree of coherence between your and your colleagues' visions for instruction and assessment?

We have asked these vision questions to more than ten thousand mathematics teachers across the United States since 2011, and the answers have been consistent: wide variance on mathematics instruction and assessment coherence from teacher to teacher (low scores of 1, 2, or 3 mostly) and general agreement that the idea of some type of a formative assessment process is supposed to be in a vision for mathematics instruction and assessment.

A favorite team exercise we use to capture the vision for instruction and assessment is to ask a team of three to five teachers to draw a circle in the middle of a sheet of poster paper. We ask each team member to write a list (outside of the circle) of three or four vital adult behaviors that reflect his or her vision for instruction and assessment. After brainstorming, the team will have twelve to fifteen vital teacher behaviors.

We then ask the team to prepare its vision for mathematics instruction and assessment inside the circle. The vision must represent the vital behaviors each team member has listed in eighteen words or less. We indicate, too, that the vision should describe a "compelling picture of the school's future that produces energy, passion, and action in yourself and others" (Kanold, 2011, p. 12).

Team members are allowed to use pictures, phrases, or complete sentences, but all together the vision cannot be more than eighteen words. In almost every case, in all of our workshops, professional development events, conferences, institutes, and onsite work, we have been asked a simple, yet complex question: *How?* How do you begin to make decisions and do your work in ways that will advance your vision for mathematics instruction and assessment in your school? How do you honor what is inside your circle? And how do you know that your circle, your defined vision for mathematics instruction and assessment, represents the "right things" to pursue that are worthy of your best energy and effort?

In our 2012 *Common Core Mathematics in a PLC at Work* series, we explain how understanding *formative assessment* as a research-affirmed *process* for student and adult learning serves as a catalyst for successful CCSS mathematics content implementation. In the series, we establish the pursuit of assessment as a process of formative feedback and learning for the students and the adults as a highly effective practice to pursue (see chapter 4 of the series).

In this handbook, we provide tools for *how* to achieve that collaborative pursuit: how to engage in ten *high-leverage team actions* (HLTAs) steeped in a commitment to a vision for mathematics instruction and assessment that will result in greater student learning than ever before.

A Cycle for Analysis and Learning: The Instructional Unit

The mathematics unit or chapter of content creates a natural cycle of manageable time for a teacher's and team's work throughout the year. What is a *unit*? For the purposes of your work in this handbook, we define a *unit* as a chunk of mathematics content. It might be a chapter from your textbook or other materials for the course, a part of a chapter or set of materials, or a combination of various short chapters or content materials. A unit generally lasts no less than two to three weeks and no more than four to five weeks.

As DuFour, DuFour, and Eaker (2008), the architects of the PLC at Work process, advise, there are four critical questions every collaborative team in a PLC at Work culture asks and answers on a unit-by-unit basis:

> 1. What do we want all students to know and be able to do? (The essential learning standards)
> 2. How will we know if they know it? (The assessment instruments and tasks teams use)
> 3. How will we respond if they don't know it? (Formative assessment processes for intervention)
> 4. How will we respond if they do know it? (Formative assessment processes for extension and enrichment)

The unit or chapter of content, then, becomes a natural cycle of time that is not too small (such as one week) and not too big (such as nine weeks) for meaningful analysis, reflection, and action by teacher teams throughout the year as they seek to answer the four critical questions of a PLC.

A mathematics unit should not last longer than three to four weeks, and at most should be analyzed based on content standard clusters—that is, three to five essential standards (or sometimes a cluster of mathematics standards) for the unit. Thus, you should monitor your grade-level or course-based teacher teams for this type of analysis about eight to twelve times per year, and an administrative central office team or district office team might do this type of analysis on a larger cycle, such as every nine weeks.

As indicated, one of your primary leadership jobs will be to know when the unit transition moments are occurring for your grade-level or course-based teams and to provide the essential time teams need to meet to do the work of ending and reflecting on one unit and beginning and preparing for the next unit throughout the school year.

This Mathematics at Work™ handbook consists of three chapters that fit the natural rhythm of your ongoing school year work as a school or district leader. The chapters bring a focus to ten high-leverage team actions that your grade-level or course-based teams take either before, during, or after a unit of instruction as they respond to the four critical questions of a PLC throughout the year. Figure I.1 lists the ten HLTAs within their time frame in relation to the unit of instruction (before, during, or after) and then links the actions to the critical questions of a PLC that they address.

High-Leverage Team Actions	1. What do we want all students to know and be able to do?	2. How will we know if they know it?	3. How will we respond if they don't know it?	4. How will we respond if they do know it?
Before-the-Unit Team Actions				
HLTA 1. Making sense of the agreed-on essential learning standards (content and practices) and pacing	Fully			
HLTA 2. Identifying higher-level-cognitive-demand mathematical tasks	Fully	Partially		
HLTA 3. Developing common assessment instruments	Partially	Fully		
HLTA 4. Developing scoring rubrics and proficiency expectations for the common assessment instruments		Partially		
HLTA 5. Planning and using common homework assignments	Partially	Fully	Partially	Partially
During-the-Unit Team Actions				
HLTA 6. Using higher-level-cognitive-demand mathematical tasks effectively	Partially	Fully		
HLTA 7. Using in-class formative assessment processes effectively	Partially	Partially	Fully	Fully
HLTA 8. Using a lesson-design process for lesson planning and collective team inquiry	Fully	Fully	Fully	Fully
After-the-Unit Team Actions				
HLTA 9. Ensuring evidence-based student goal setting and action for the next unit of study			Fully	Fully
HLTA 10. Ensuring evidence-based adult goal setting and action for the next unit of study			Fully	Fully

[gray box] = Fully addressed with high-leverage team action

[half-gray box] = Partially addressed with high-leverage team action

Figure I.1: High-leverage team actions aligned to the four critical questions of a PLC.

Visit **go.solution-tree.com/mathematicsatwork** to download a reproducible version of this figure.

Before the Unit

In chapter 1, we provide insight into the work of your collaborative teams *before* the unit begins, along with the tools you need in this phase. Your collaborative teams should (as best they can) complete this teaching and assessing work in preparation for the unit.

There are five before-the-unit high-leverage team actions for collaborative team agreement on a unit-by-unit basis.

> HLTA 1. Making sense of the agreed-on essential learning standards (content and practices) and pacing
>
> HLTA 2. Identifying higher-level-cognitive-demand mathematical tasks
>
> HLTA 3. Developing common assessment instruments
>
> HLTA 4. Developing scoring rubrics and proficiency expectations for the common assessment instruments
>
> HLTA 5. Planning and using common homework assignments

Once you and your teams have taken these action steps, the mathematics unit begins.

During the Unit

In chapter 2, we provide the tools for and insight into the formative assessment work of your collaborative teams *during* the unit. This chapter emphasizes deeper understanding of content, discussing the Common Core Mathematical Practices and processes and using higher-level-cognitive-demand mathematical tasks effectively. It helps your teams with daily lesson design as ongoing in-class student instruction becomes part of a teacher-led and student-engaged formative process.

This chapter introduces three during-the-unit high-leverage team actions teams work through on a daily basis.

> HLTA 6. Using higher-level-cognitive-demand mathematical tasks effectively
>
> HLTA 7. Using in-class formative assessment processes effectively
>
> HLTA 8. Using a lesson-design process for lesson planning and collective team inquiry

The end of each unit results in some type of student assessment. Your teachers pass back the assessments scored and with feedback. Then what? What are students to do? What are teacher team members to do?

After the Unit

What is happening in your school or across schools in the district when a unit of mathematics ends for each grade level or mathematics course?

After instruction for a unit is over and teachers have given the common assessment, students should be expected to reflect on the results of their work and take action on the formative feedback from the mathematics unit assessment instrument to advance their learning of the essential standards. Your teacher teams need to establish a culture in which students welcome error as an opportunity to learn. You will need to monitor this process to ensure it is happening with all of your grade-level or course-based teams.

In addition, there is another primary formative purpose to using a common end-of-unit assessment, which Hattie (2012) describes in *Visible Learning for Teachers*: "This [teacher collaboration] is not critical reflection, but critical reflection *in light of evidence* about their teaching" (p. 19, emphasis added).

From a practical point of view, the collaborative teams best serve the act of "reflection in light of evidence" by performing an end-of-unit analysis of students' common assessment results. Then through this analysis, you help teams focus their teaching and assessing actions for the next unit of student learning and help revise their instructional plans in advance of teaching the unit next year.

Thus, there are two end-of-unit high-leverage team actions for reflection and action that teams work through on a unit-by-unit basis.

> HLTA 9. Ensuring evidence-based student goal setting and action for the next unit of study
>
> HLTA 10. Ensuring evidence-based adult goal setting and action for the next unit of study

In *Principles to Actions: Ensuring Mathematical Success for All*, NCTM (2014) presents a modern-day view of professional development for mathematics teachers: building the knowledge capacity of every teacher. More importantly, however, you must help your teachers to intentionally *act* on that knowledge and transfer what they learn into daily classroom practice. The ten high-leverage team actions we present in this handbook are one way to do so. For more information on the connection between these two documents, see appendix E (page 143).

Although given less attention, the difficult work of collective inquiry and action orientation and experimentation has a more direct impact on student learning than when you allow teachers to work in isolation. Collective inquiry and action orientation and experimentation may also contribute to a reduction in traditional achievement gaps (Hattie, 2009; Moller, Mickelson, Stearns, Banerjee, & Bottia, 2013). It is through your commitment to guiding your teacher teams through the PLC at Work process that teacher inquiry and experimentation will flourish and teachers will find meaning in their collaborative work with colleagues.

In *Great by Choice*, Jim Collins (Collins & Hansen, 2011) asks, "Do we really believe that our actions count for little, that those who create something great are merely lucky, that our circumstances imprison us?" He then answers, "Our research stands firmly against this view. Greatness is not primarily a matter of circumstance; greatness is first and foremost a matter of conscious choice and discipline" (p. 181). We hope this handbook helps you to focus your leadership time, energy, choices, and pursuit of a great teaching journey for every teacher.

CHAPTER 1

Before the Unit

Teacher: Know thy impact.

—John Hattie

As a school leader, you are and always will be a teacher—of adults. Thus, the Hattie quote that opens this chapter is for you too. What will be your impact on the adults in your school or district, every month, every day, and on every unit of instruction? The ultimate outcome of before-the-unit planning is for your teachers to develop a clear understanding of the shared expectations for student learning during the unit. Do you expect a teacher learning culture that understands mathematics as an effort-based and not an ability-based discipline? Do you have high expectations that every teacher can ensure all students learn?

Your collaborative teams, in conjunction with district mathematics curriculum team leaders, prepare a roadmap that describes the knowledge students will know and be able to demonstrate at the conclusion of the unit. To create this roadmap, each collaborative team prepares and organizes work around five before-the-unit high-leverage team actions that you will need to monitor.

> HLTA 1. Making sense of the agreed-on essential learning standards (content and practices) and pacing
>
> HLTA 2. Identifying higher-level-cognitive-demand mathematical tasks
>
> HLTA 3. Developing common assessment instruments
>
> HLTA 4. Developing scoring rubrics and proficiency expectations for the common assessment instruments
>
> HLTA 5. Planning and using common homework assignments

These five team pursuits are based on step one of the PLC teaching-assessing-learning cycle (Kanold, Kanold, & Larson, 2012) shown in figure 1.1 (page 10). This cycle drives your pursuit of a meaningful formative assessment and learning process for your teacher teams and for your students throughout each unit of instruction during the year.

In this chapter, we describe each of the five before-the-unit teacher team actions in more detail (the what) along with suggestions for how to achieve these pursuits (the how). Each HLTA section ends with an opportunity for you to evaluate the current reality for your teams (team progress). The chapter ends with time for reflection and action (setting your Mathematics at Work priorities for team action).

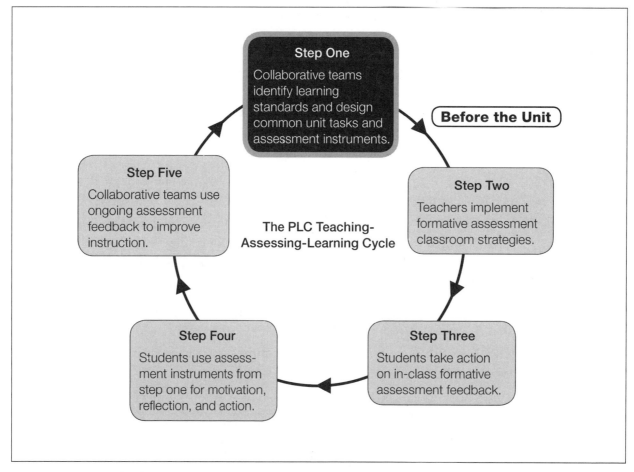

Source: Kanold, Kanold, & Larson, 2012.

Figure 1.1: Step one of the PLC teaching-assessing-learning cycle.

HLTA 1: Making Sense of the Agreed-On Essential Learning Standards (Content and Practices) and Pacing

> *An excellent mathematics program includes curriculum that develops important mathematics along coherent learning progressions and develops connections among areas of mathematical study and between mathematics and the real world.*
>
> —National Council of Teachers of Mathematics

For most grade levels (K–8) or courses (9–12) there will be eight to twelve mathematics units (or chapters or modules) during the school year. How do you help your collaborative teams develop their understanding for each of the agreed-on and essential mathematics learning standards for the units throughout the year—especially when the mathematics content may not be your area of expertise? You can help by asking for and monitoring several key collaborative team artifacts.

Recall there are four critical questions every collaborative team in a PLC asks and answers on an ongoing unit-by-unit basis.

1. What do we want all students to know and be able to do? (The essential learning standards)
2. How will we know if they know it? (The assessment instruments and tasks teams use)
3. How will we respond if they don't know it? (Formative assessment processes for intervention)
4. How will we respond if they do know it? (Formative assessment processes for extension and enrichment)

High-Leverage Team Action	1. What do we want all students to know and be able to do?	2. How will we know if they know it?	3. How will we respond if they don't know it?	4. How will we respond if they do know it?
Before-the-Unit Action				
HLTA 1. Making sense of the agreed-on essential learning standards (content and practices) and pacing	■			

■ = Fully addressed with high-leverage team action

This first high-leverage team action enhances teacher clarity on the first PLC critical question: What do we want all students to know and be able to do?

The What

———————————————————— **Real-Life Leadership Scenario** ————————————————————

We were working with a grade-level team on the essential standards for the unit teachers were about to teach. In an offhand remark to the team, I [Tim] said, "Well, of course everyone on the team teaches to the same standards, right? This became expected practice back in the late 1980s." A few team members started to giggle. When I asked them what was funny, they indicated that there was one member who rarely taught the expected standards and that she wasn't there today. I asked, "Well, how is she allowed to do that?" They answered that it had always been that way; she was just different. I asked if they saw the inequity this caused for students, and they indicated that yes, they did, but the group became uncomfortable, so we moved on. I made a note to talk to the principal about the matter as soon as possible. I wondered, too, what the principal would actually do to help the team resolve this inequity in student learning.

The essential learning standards for the unit—the guaranteed and viable mathematics curriculum—include *what* (content standard clusters and standards) students will learn, *when* they will learn it (the pacing of the unit), and *how* they will learn it (often via process standards such as the Common Core Standards for Mathematical Practice or other state-designed process standards). The Common Core Standards for Mathematical Practice "describe varieties of expertise that mathematics educators at all levels should seek to develop in their students" (National Governors Association Center for Best Practices [NGA] & Council of Chief State School Officers [CCSSO], 2010, p. 6). Following are eight Standards for Mathematical Practice, which we include in full in appendix A (page 131).

1. Make sense of problems and persevere in solving them.
2. Reason abstractly and quantitatively.
3. Construct viable arguments and critique the reasoning of others.
4. Model with mathematics.
5. Use appropriate tools strategically.
6. Attend to precision.
7. Look for and make use of structure.
8. Look for and express regularity in repeated reasoning. (NGA & CCSSO, 2010, pp. 6–8)

Every teacher and teacher team you lead must be tight about teaching to the essential learning (content and process) standards for the unit. The answer to the first critical question (What do we want all students to know and be able to do?) is not optional or a choice for each teacher to make individually. Your role, then, is to monitor and ensure every course-based or grade-level team member receives the professional development, instruction, engagement, and confidence necessary to teach each essential standard for the unit well.

A unit of mathematics instruction connects topics in mathematics that are naturally grouped together—the essential ideas or content standard clusters. Each of those units (sometimes called *chapters*, *modules*, or both depending on your state, district, or curriculum) should consist of about four to six essential learning standards every student of the course is expected to learn. This is considered the optimal grain size for all of your assessment and analysis work as a teacher team. Too many standards per unit become overwhelming for teachers. Too few, and teachers cannot focus student learning and their support. Four to six total essential learning standards for the unit are just right.

> A simple step would be to ask the team to show you *how* this is being done. Ask your teams to show you the number of essential learning standards they are teaching for a current unit. Does the grain size of their focus make sense to you? If not, advise them to revise and adjust the number of essential learning standards for the unit.

Do all team members agree on the essential learning standards, and do you have evidence they are teaching to those standards? If not, be clear about your essential learning standard expectations for *all* teachers on the team.

Although the essential learning standards for the unit might be developed as part of a district curriculum task force, your collaborative teams should take time during lesson-design discussions to make sense of the essential standards (for example, you could always ask a team to explain to you how it is making sense of an essential standard for the unit and to provide you with a mathematics task or example that represents that standard).

As your collaborative teams unpack the content standards (or your district's essential learning standards), it is also important for them to decide which Mathematical Practices or processes make mathematical sense to engage students throughout the unit of instruction and as part of each lesson. *Unpacking*, in this case, means making sense of the mathematics listed in the standard, making sense of how the content connects to content learned in other mathematics courses as well as within the current course, and making sense of how students might develop both conceptual understanding and procedural skill with the mathematics listed in the standard.

> Thus, throughout the year, you can always ask a teacher or team, "For today's mathematics lesson, can you explain to me the essential learning standard *and* the essential Mathematical Practices or process standard I should expect to see the students doing? How will students demonstrate understanding?"

More detail on implementing Mathematical Practices and processes as part of lesson design is provided in HLTA 6 on page 63.

The How

How can you help your teams unpack the content and link it to student practices for any unit, especially if mathematics is not your core discipline of understanding? You can help your teacher teams by providing them with the time necessary for deep discussions about the meaning of the essential learning standards *before* the unit begins and by actively monitoring the results of those team discussions.

Teams Make Sense of the Essential Learning Standards

A key outcome of this first high-leverage team action is for your teachers to personally and collaboratively make sense of the essential learning standards with an eye toward planning for student engagement in the Mathematical Practices that support them. This needs to occur *before* the unit begins for teachers to take full advantage of their instructional time and effort during the unit.

At Stevenson High School District 125, a PLC school district, school leaders (including the lead author of this handbook, Tim Kanold, who was the director of mathematics and the superintendent) were required to ask the teachers for evidence that they were using national resources in their work. They needed to address how national thought leaders and expert organizations were informing local practice. Your collaborative teams may need your help securing outside resources to make sense of the mathematics involved in the learning standards within a unit. Reading the background information in your school textbook or digital teacher's editions can be a good source for this foundational knowledge, as can resources from the National Council of Teachers of Mathematics (www.nctm.org), such as the *Essential Understanding* series.

> In general, you can monitor and support HLTA 1 by asking each team to use figure 1.2 as a discussion tool during the year as teachers break down the major essential learning standards for student understanding in each unit.

This is a form you can ask teams to complete for each unit and *turn in to you for your review.*

However, there is one caveat: you must review the completed form and provide feedback (good or bad) to the teacher team. Otherwise, the team will view completion of the form as a perfunctory exercise and not take it seriously.

Figure 1.2 offers one way for you to collect evidence of your teams' discussions on this important high-leverage team action. When unpacking the essential learning standards through these types of questions, your collaborative teams develop understanding of the essential learning standards, including daily learning objectives, the prerequisite knowledge and vocabulary needed, and the appropriate Mathematical Practices to support student learning.

Teams Make Sense of the Content Progressions for the Unit

You should also ask each team to provide you with a unit calendar that shows the essential learning standard progression of the unit. Figure 1.3 (pages 16–17) provides a sample eighth-grade geometry unit. By viewing the calendar, you know the approximate pacing of the unit, how many days (lessons) are spent on each essential learning standard, when the unit ends, interventions you might need to support during and after the unit, and questions that might arise from the broader school community, including parents.

Directions: Work with your collaborative team to unpack the essential learning standards and the associated Mathematical Practices relevant to the current unit of study for your grade level or course.

Teacher team: _____

Name of unit: _____

1. List the agreed-on essential learning standards for this unit.

2. What is the prerequisite knowledge needed to engage students with each essential learning standard?

3. What is the time frame available to teach this unit, and how will that time be distributed for each essential learning standard?

4. What are the mathematics vocabulary and literacy skills necessary for student success in this unit?

5. What are specific teaching strategies, tasks, and tools that will most effectively support each essential learning standard for the unit?

6. Which Mathematical Practices or processes should be highlighted during the unit in order to better engage students in the process of understanding each learning standard?

7. Identify the specific lessons that will highlight mathematical modeling for the standards of the unit.

Figure 1.2: Discussion tool for making sense of the agreed-on essential learning standards for the unit.

Visit **go.solution-tree.com/mathematicsatwork** to download a reproducible version of this figure.

Unit Plan: Twenty Instructional Days

Day 1	Day 2	Day 3	Day 4	Day 5
8.G.1:	**8.G.1:**	**8.G.1:**	**8.G.1:**	**8.G.2:**
I can verify experimentally the properties of rotations, reflections, and translations. Exploration Using Geometry Software Students will manipulate shapes through translations and reflections to make conjectures about their observations.	I can verify experimentally the properties of rotations, reflections, and translations. Exploration Using Geometry Software Students will manipulate shapes through rotations to make conjectures about their observations.	I can verify experimentally the properties of rotations, reflections, and translations. Informal Assessment Students create a transformation to prove the properties of transformations.	I can verify experimentally the properties of rotations, reflections, and translations. Students will engage in a final vocabulary activity using the Frayer Model to finalize understanding about the properties of transformations.	I can demonstrate the congruence of two-dimensional figures using the properties of rotations, reflections, and translations. Students will discuss congruence and begin creating congruent figures through translations.

Day 6	Day 7	Day 8	Day 9	Day 10
8.G.2:	**8.G.2:**	**8.G.2:**	**8.G.2:**	**8.G.2:**
I can demonstrate the congruence of two-dimensional figures using the properties of rotations, reflections, and translations. Students will explore congruent figures that are reflected over the x-axis and y-axis.	I can demonstrate the congruence of two-dimensional figures using the properties of rotations, reflections, and translations. Students will explore congruent figures that are reflected over other lines.	I can demonstrate the congruence of two-dimensional figures using the properties of rotations, reflections, and translations. Students will explore rotations of shapes and identify if they are congruent.	I can demonstrate the congruence of two-dimensional figures using the properties of rotations, reflections, and translations. Students will create rotations that result in congruent figures and noncongruent figures using physical models and geometry software.	Informal Assessment I can demonstrate the congruence of two-dimensional figures using the properties of rotations, reflections, and translations. Students will create a sequence of transformations for a peer to decide if they are congruent.

continued ↓

Day 11	Day 12	Day 13	Day 14	Day 15
8.G.3:	**8.G.3:**	**8.G.4:**	**8.G.4:**	**8.G.4:**
I can describe the effect of translations, rotations, and reflections on two-dimensional figures using coordinates. Students will apply their knowledge of transformations and learn how to use coordinates to describe a transformation or a series of transformations.	I can describe the effect of translations, rotations, and reflections on two-dimensional figures using coordinates. Students will apply their knowledge of transformations and learn how to use coordinates to describe a transformation or a series of transformations.	I can demonstrate that two figures are similar by using the properties of ~~dilations,~~ rotations, reflections, and translations of two-dimensional figures. Students will begin to explore the meaning of similar figures and the difference between similar and congruent.	I can demonstrate that two figures are similar by using the properties of ~~dilations,~~ ~~rotations,~~ reflections, and translations of two-dimensional figures. Students will discuss how two figures can be similar using reflections and translations. They will examine examples and nonexamples.	I can demonstrate that two figures are similar by using the properties of ~~dilations,~~ rotations, reflections, and translations of two-dimensional figures. Students will discuss how two figures can be similar using rotations, reflections, and translations. They will examine examples and nonexamples.

Day 16	Day 17	Day 18	Day 19	Day 20
8.G.3 (Part 2):	**8.G.3 (Part 2) and 8.G.4:**	**8.G.4:**	**Review for Unit 1**	**Assessment for Unit 1**
I can describe the effect of dilations on two-dimensional figures using coordinates. Students will explore the effect of dilations on coordinates for two-dimensional figures using models and geometry software. Students will establish generalizations about effects.	I can describe the effect of dilations on two-dimensional figures using coordinates. Students will continue building their knowledge of dilations and how to represent the effect of a dilation using coordinates through various tasks.	I can describe a sequence of transformations between two figures that exhibits the similarity between them. Students will be given and will create a sequence of transformations between two figures and describe the sequence.	Students will combine all standards together.	

Notes for Unit 1

When working through each standard, we may not need to break up the learning targets by each transformation; however, it may also help students to take an in-depth look at each transformation. This is something we will monitor throughout the unit and make notes on for next year. Also, 8.G.1 will continue to be embedded throughout instruction in this unit. Before moving on to similarity, we will ensure all students have a solid understanding of congruence and how it relates to transformations. Most work and dialogue during this unit will occur in teams of four. Students will present their thinking and listen to the thinking and reasoning of others to fully develop their understanding and their demonstration for the overarching unit, Mathematical Practices 1, "Make sense of problems and persevere in solving them," and 4, "Model with mathematics."

Note: Crossed-out text indicates that only a certain portion of the standard is the focus.
Source: Adapted with permission from Aptakisic-Tripp CCSD 102, Buffalo Grove, Illinois.
Source for standards: NGA & CCSSO, 2010, pp. 55–56.

Figure 1.3: Sample unit progression for eighth-grade geometry.

Visit **go.solution-tree.com/mathematicsatwork** to download a reproducible version of this figure.

This first high-leverage team action is both a district *and* a teacher team responsibility. The district office does need to provide guidance to principals, leaders, and teachers as to the proper scope and sequence of the essential learning standards of the unit to ensure a guaranteed and viable curriculum that addresses student mobility issues.

> You can't take away from team members the real work they must do; if you take away the work, you take away the learning. Your course- or grade-based team members need to be clear on the intent of the essential learning standards, the rationale for teaching the standards in a specific order, and the nuances of the meaning and intent of each essential standard.

At a minimum, you can expect your teams to write a set of unit notes using the questions provided in figure 1.2 on page 15 (and turn those notes in to you) as they work to better understand the intent of the mathematics content, the mathematics content progressions, and the overarching Standards for Mathematical Practice for that unit. Make sure you are connected to the electronic posting of all unit notes.

Team Progress

It is helpful to diagnose the collaborative team reality and action prior to launching any unit. You can ask each team to assess its progress for HLTA 1, making sense of the agreed-on essential learning standards (content and practices) and pacing, by using table 1.1. It matters less which stage your teams are at and more that team members are committed to working together to focus on understanding the essential learning standards and the best mathematical tasks and strategies for increasing student understanding and achievement as teams seek stage IV—sustaining.

As your teams unpack the essential learning standards, they also need to identify and prepare for higher-level-cognitive-demand mathematical tasks related to those essential learning standards. It is necessary for them to include tasks at varying levels of demand during instruction. The idea is to match the tasks and their cognitive demand to the essential learning standard expectations for the unit. Selecting mathematical tasks together is the topic of the second high-leverage team action, HLTA 2.

Table 1.1: Before-the-Unit-Begins Status Check Tool for HLTA 1—Making Sense of the Agreed-On Essential Learning Standards (Content and Practices) and Pacing

Directions: Discuss your perception of your team's progress on the first high-leverage team action—making sense of the agreed-on essential learning standards (content and practices) and pacing. Defend your reasoning.

Stage I: Pre-Initiating	Stage II: Initiating	Stage III: Developing	Stage IV: Sustaining
We do not discuss the essential learning standards of the unit prior to teaching it.	We discuss and reach agreement on the four to six essential learning standards for the unit.	We unpack the intent of each essential learning standard for the unit and discuss daily learning objectives to achieve each essential standard.	We connect the four to six essential learning standards to the Mathematical Practices before the unit begins.
We do not know which essential learning standards other colleagues of the same course or grade level teach during the unit.	We discuss and share how to develop student understanding of the essential learning standards during the unit.	We collaborate with our colleagues to make informed decisions about instruction of the essential learning standards for each lesson in the unit.	We have procedures in place to review the effectiveness of the students' roles, activities, experiences, and success on the essential learning standards during the unit.
We do not discuss lesson tasks.	We connect and align some lesson tasks to the essential learning standards for the unit.	We share effective teaching strategies for the essential learning standards of the unit.	We have procedures in place that ensure our team aligns the most effective mathematical tasks and instructional strategies to the content progression established in our overall unit plan components.
We do not discuss Mathematical Practices and processes as part of our unit planning.	We discuss Mathematical Practices and processes that best align to the essential learning standards for the unit.	We agree on Mathematical Practices and processes that best align to the learning standards for the unit.	We implement Mathematical Practices and processes that best align to the learning standards for the unit.

Visit **go.solution-tree.com/mathematicsatwork** to download a reproducible version of this table.

HLTA 2: Identifying Higher-Level-Cognitive-Demand Mathematical Tasks

The function of education is to teach one to think intensively and to think critically.

—Martin Luther King Jr.

The mathematical tasks and activities your teachers and teacher teams choose to use each day and for every unit are the backbone for effective student learning. If you do not monitor those choices, they can often be a source of deep inequity in your school or district. The daily math problems or tasks your teachers use for student learning should not be made in isolation from their colleagues. The daily mathematical tasks chosen by your teacher teams each day represent how they answer the first critical question of a PLC: *What do we want all students to know and be able to do?*

The mathematical tasks and activities teachers choose each day also partially answer the second PLC at Work critical question for collaborative teams, How will we know if they know it? The nature of the mathematical tasks your teachers choose—higher- or lower-level cognitive demand—is as essential to student learning as the content and the process standards chosen for the unit.

High-Leverage Team Action	1. What do we want all students to know and be able to do?	2. How will we know if they know it?	3. How will we respond if they don't know it?	4. How will we respond if they do know it?
Before-the-Unit Action				
HLTA 2. Identifying higher-level-cognitive-demand mathematical tasks	▨	▨□		

▨ = Fully addressed with high-leverage team action

▨□ = Partially addressed with high-leverage team action

The What

————————————— **Real-Life Leadership Scenario** —————————————

I [Tim] was meeting with a high school geometry team in Southern California. At the team meeting, we were taking a close look at the type of mathematics problems (tasks) teachers would be using in class the next day. This being a geometry class, working on geometry standards, I was expecting some pretty exciting student investigations through the use of some higher-level-cognitive-demand tasks. I asked the teacher team two very simple questions.

1. *"How do you decide each day which mathematical tasks you will use to teach the standards for the lesson?"*

2. *"How will you ensure there is a balance of higher- and lower-level-cognitive-demand tasks used for student exploration and learning?"*

The teachers responded, "What do you mean by mathematical task and cognitive demand? We use the mathematics problems like the ones in our book, so our students will be ready for homework. Our students don't really do the harder stuff; we are mostly trying to get them through the course. You will understand better after you hang around the school for the next few months. This is a pretty tough neighborhood."

I asked the teachers on the team if they were getting help from their department chair, and one of the teachers said she was the department chair, and mentioned to me it was her suggestion to focus only on what the students could do and not that harder stuff. I took a deep breath and dove in wondering to whom in the administration I could go next to guide the teacher team and the entire department toward a more student-engaged approach via the tasks chosen. If it wasn't to be the department chair, then who? Who could the principal turn to for access to Hattie's (2012) research summary that supports teaching problem solving as a way to enhance basic skill development in students?

What is a mathematical task?

NCTM first identified the term *mathematical task* in its (1991, 2007) *Professional Teaching Standards* as "worthwhile mathematical tasks" (p. 24). Melissa Boston and Peg Smith (2009) later provide this succinct definition: "A mathematical task is a single complex problem or a set of problems that focuses students' attention on a specific mathematical idea" (p. 136).

Mathematical tasks include activities, examples, or problems to complete as a whole class, in small groups, or individually. The tasks provide the rigor (levels of complex reasoning from the conceptual understanding, procedural fluency, and application of the tasks) that students require and thus become an essential aspect of your teams' collaboration and discussion. In short, the tasks are the problems your teachers choose to determine the pathway of student learning and to assess student success along that pathway. Your teachers are empowered to decide what and how a student learns through their choice and the use of the mathematical tasks and activities that students experience. As Glenda Lappan and Diane Briars (1995) state:

> There is no decision that teachers make that has a greater impact on students' opportunities to learn and on their perceptions about what mathematics is than the selection or creation of the tasks with which the teacher engages students in studying mathematics. (p. 139)

The selection of worthwhile mathematical tasks is so critical that it is one of the eight research-informed instructional strategies listed in *Principles to Actions* (NCTM, 2014). A key collaborative team decision, then, is to decide which tasks to use in a particular lesson or unit to help students attain the essential learning standards. A growing body of research links students' engagement in higher-level-cognitive-demand tasks to overall increases in mathematics learning, not just in the ability to solve problems (Hattie, 2012; Resnick, 2006).

Higher-level-cognitive-demand lessons or tasks are those that provide "opportunities for students to explain, describe, justify, compare, or assess; to make decisions and choices; to plan and formulate questions; to exhibit creativity; and to work with more than one representation in a meaningful way" (Silver, 2010, p. 2). In contrast, lessons or tasks with only lower-level cognitive demand are "characterized as opportunities for students to demonstrate routine applications of known procedures or to work with a complex assembly of routine subtasks or non-mathematical activities" (Silver, 2010, p. 2).

Take caution: selecting a task with higher-level cognitive demand does not ensure students will engage in rigorous mathematical activity (Jackson, Garrison, Wilson, Gibbons, & Shahan, 2013). The cognitive demand of a mathematical task is often lowered (perhaps unintentionally) during the implementation phase of the lesson (Stein, Remillard, & Smith, 2007). Thus, during the planning phase before the unit begins, your teams should discuss how they would respond when students urge them to lower the cognitive demand of the task during the lesson. Supporting productive struggle in learning mathematics is one of the eight research-informed mathematics teaching practices outlined in *Principles to Actions* (NCTM, 2014). Strategies to avoid cognitive decline during task implementation are discussed further in chapter 2 (page 63, HLTA 6).

> You can help your teacher teams by asking them to respond to several mathematical task questions before each unit begins and then turn in those responses to you. Remember, a major part of your work is to know what the teacher teams are doing, and to take action that both validates (supports) their work and provides feedback for improvement (accountability) as needed.
>
> 1. How do we define and differentiate between higher-level-cognitive-demand and lower-level-cognitive-demand tasks for each essential standard of the unit?
>
> 2. How do we select common higher-level-cognitive-demand and lower-level-cognitive-demand tasks for each essential standard of the unit?
>
> 3. How do we create higher-level-cognitive-demand tasks from lower-level-cognitive-demand tasks for each essential standard of the unit?
>
> 4. How do we use and apply higher-level-cognitive-demand tasks for each essential standard during the unit?
>
> 5. How will we respond when students urge us to lower the cognitive demand of the task during the implementation phase of the lesson?

Visit **go.solution-tree.com/mathematicsatwork** to download these questions as a discussion tool.

You can also support your teacher teams in this work by expecting them to define, identify, and create higher-level- and lower-level-cognitive-demand tasks and to prepare for the use of higher-level-cognitive-demand tasks during each unit.

The How

A critical step in selecting, planning, and eventually using higher-level-cognitive-demand mathematical tasks in class is for your collaborative teams to use designated team time to work the task before giving it to students. Working the task together should be an expected agenda item as each team provides information about possible solution strategies or pathways that students might demonstrate. When you take a look at the minutes or agendas of your teams' meetings, be sure to look for evidence of discussions about the math tasks they are using in class.

Defining Higher- and Lower-Level-Cognitive-Demand Mathematical Tasks

Your teachers choose mathematical tasks for every lesson, every day. Take a moment to describe how they do this part of their work. Do they make those decisions alone, with team members, before the unit begins, or the night before they teach the lesson? Where do they locate the mathematical tasks? From the textbook? Online? From district resources? Sit in on a team meeting, and ask them these questions.

Most importantly, how would you describe the rigor of the mathematical tasks your teachers are choosing for your students? Rigor is not whether a mathematical task or problem is considered hard. For example, "Find the tangent $5\pi/6$," "Find 2^{-3}," or "What is 6×7" might be hard for some, but they are not rigorous tasks. *Rigor of a mathematical task* is defined in this handbook as the level and the complexity of reasoning the student requires during the task (Kanold, Briars, & Fennell, 2012). A more rigorous version of the latter task might be something like, "Provide two different ways to determine 6×7 using facts you might know." There are several ways to label the demand or rigor of a task; however, for the purposes of this handbook, tasks are classified as either lower-level cognitive demand or higher-level cognitive demand as defined by Smith and Stein (1998) in their task-analysis guide and printed in full as appendix C (page 139). *Lower-level-cognitive-demand tasks* are typically focused on memorization or performing standard or rote procedures without attention to the properties that support those procedures (Smith & Stein, 2011).

Higher-level-cognitive-demand tasks are tasks for which students do not have a set of predetermined procedures to follow to reach resolution or, if the tasks involve procedures, they require that students provide the justification for why and how the procedures can be performed. Smith and Stein (2011) describe these procedures as "procedures with connections" (p. 16) as opposed to "procedures without connections," the designation they use for lower-level-cognitive-demand tasks that are not just based on memorization.

Figure 1.4 shows the two levels of cognitive demand and the four categories within the levels.

Lower-Level Cognitive Demand

Memorization: Requires eliciting information such as a fact, definition, term, or a simple procedure, as well as performing a simple algorithm or applying a formula.

Procedures without connections: Requires the engagement of some mental processing beyond a recall of information.

Higher-Level Cognitive Demand

Procedures with connections: Requires complex reasoning, planning, using evidence, and explanations of thinking.

Doing mathematics: Requires complex reasoning, planning, developing, and thinking most likely over an extended period of time.

Source: Smith & Stein, 2012.

Figure 1.4: Four categories of cognitive demand.

Visit **go.solution-tree.com/mathematicsatwork** to download a reproducible version of this figure.

In all likelihood, the majority of the daily mathematical tasks your teachers choose are of the lower-level-cognitive-demand variety. You should find out if this is true for your teacher teams or not. For many teachers, there is a natural drift toward lower-level mathematical tasks. Lower-level-cognitive-demand tasks take less time in class and do not require much complex reasoning by students. They often contain one most efficient pathway and do not require student demonstration of multiple solution pathways. The efficiency of lower-level-cognitive-demand tasks is appealing. They are much easier to manage in class as a general rule and easily serve direct instruction from the front of the room. Mathematical content and practice standard expectations require students to demonstrate *understanding*. This requires a shift to a more *balanced* task approach during the unit—the use of both higher- and lower-level-cognitive-demand tasks. In most classrooms, this will require an increase in the use of higher-level-cognitive-demand tasks.

You can use figure 1.5, an example of a higher-level-cognitive-demand task, to help your teachers demonstrate understanding of the criteria for higher-level cognitive demand. Use the four questions in the task as a guide.

Standard: I can understand the meaning of equivalent expressions.

Look at each expression. Is it equivalent to $\dfrac{x + 3y}{2}$?

Select Yes or No for expressions A–D.

A. $\dfrac{4x + 3y}{8}$ ○ Yes   ○ No

B. $\dfrac{5}{4}\left(\dfrac{2x + 6}{5}\right)$ ○ Yes   ○ No

C. $\dfrac{1}{2}(x + 3y)$ ○ Yes   ○ No

D. $\dfrac{2}{3}\left(\dfrac{5x}{6} + \dfrac{9y}{4} - \dfrac{x}{12}\right)$ ○ Yes   ○ No

Explain why each choice (A, B, C, and D) is equivalent or why it is not equivalent.

Directions: Find a solution pathway to the problem by yourself first, and then discuss the mathematics task with your collaborative team.

1. How are your collaborative team members' responses the same? How do they differ?

2. How does this task (and your solution pathway to the task) support the essential learning standard for equivalent expressions, and what is the prerequisite knowledge needed for the task?

3. How does this task meet the criteria for higher-level cognitive demand?

4. Which Mathematical Practices or processes might students engage while solving this higher-level-cognitive-demand mathematical task?

5. Where might students get stuck when trying to work on this task together?

Source for the task: Smarter Balanced Assessment Consortium, 2013a. Used with permission.

Figure 1.5: Sample higher-level-cognitive-demand task discussion tool.

Visit **go.solution-tree.com/mathematicsatwork** to download a reproducible version of this figure.

Mathematical tasks such as those described in figure 1.5 determine student understanding and reasoning. Good higher-level-cognitive-demand mathematical tasks will contain elements such as multiple entry points to the tasks (lots of ways a student could enter into the problem), multiple solution pathways (lots of ways a student could reason to a solution), and require more complex levels of reasoning beyond just finding the one "right" answer. This is the nature of good higher-level-cognitive-demand mathematical tasks.

Identifying the Cognitive Demand of the Daily Mathematical Tasks Teachers Use in Your School

Ask your grade-level or course-based teacher teams to provide the percentage of higher- to lower-level-cognitive-demand tasks they present to their students throughout the unit, along with examples of these for the unit's essential standards. You should expect teachers to roughly use a ratio of 3:1 lower-level-cognitive-demand tasks to higher-level-cognitive-demand tasks for each lesson during the unit. If this is the case for each daily lesson, it should at least be the ratio used for the balance of the unit.

As a first step in understanding the balance of current cognitive-demand levels of the mathematical tasks teachers use each day, ask your teachers to complete figure 1.6 for at least two standards of a future unit.

Name of the Unit:
For at least two of the essential standards in this unit, provide samples of the types of mathematical tasks students will experience in class, for homework, or on assessments.
Directions: Sort every task you use into the following four categories.

Lower-Level Tasks	Higher-Level Tasks
Memorization	**Procedures With Connections**
Procedures Without Connections	**Doing Mathematics**

Figure 1.6: Tool for sorting unit tasks by cognitive-demand level.

Visit **go.solution-tree.com/mathematicsatwork** to download a reproducible version of this figure.

One way you can often judge whether a mathematical task is of higher cognitive demand (or not), even if you have a limited mathematics background, is to examine the differences in what a student is expected *to do* during the task.

Since student work should be balanced with respect to the level of cognitive demand across tasks, it is important for teachers to identify expected levels of cognitive demand and ultimately adapt or create tasks for each essential learning standard in the unit as the standards progress over time.

Creating Higher-Level-Cognitive-Demand Tasks

Higher-level-cognitive-demand tasks are essential for improving student achievement in mathematics. There are many resources online and in print that can provide examples of higher-level-cognitive-demand tasks for use in class. See appendix D (page 141), go to nctm.org, or visit **go.solution-tree.com /mathematicsatwork** for a list of resources.

However, you should expect each of your teams to also learn how to *create mathematical tasks* of varying cognitive demand together (remember if you take away the *doing* of this work from the team, you take away teachers' learning). This will empower greater ownership and understanding of mathematical task design and selection by the teachers on the team.

There are several strategies you can use to change a lower-level-cognitive-demand mathematical task to higher-level cognitive demand. You can ask your teams to use the strategies in figure 1.7 to adjust a mathematical task from lower-level cognitive demand to higher-level cognitive demand.

1. Use comparison questions. (When is one situation greater than, equal to, or less than another?)

2. Ask a question across multiple representations in a task.

3. Validate a solution pathway or approach.

4. Require students to provide justifications for (explain) their solutions.

5. Evaluate the error or reasoning in a student solution and provide a correct solution pathway.

6. Create a context. Ask students to write a word problem that creates a context for the given information.

7. Ask students to determine an expression to represent a situation.

8. Create an open-ended debate-type task, so that multiple student responses will satisfy a solution to the mathematical task.

Figure 1.7: Strategies for increasing the cognitive demand of tasks.

Visit **go.solution-tree.com/mathematicsatwork** to download a reproducible version of this figure.

You should expect your teams to use the discussion tool in figure 1.8 often. It will help them to become efficient with the process of increasing the cognitive demand of the tasks they chose for each unit. From time to time, you can ask them to complete this form and turn it in to you at the end of their meeting.

Essential learning standard for the unit:

Find one lower-level-cognitive-demand task you have used for this essential standard in the past. Using one of the task-modification strategies from figure 1.7, rewrite the task at a higher level of cognitive demand. List both the lower- and the higher-level-cognitive-demand task below.

Lower Level	Higher Level

Justify the cognitive-demand level of each task above, and then prepare to discuss with your team.

Answer the following questions with your collaborative team.

1. How might what you learn about your students' understanding of the essential learning standard differ depending on the cognitive demand of the task you use during instruction?

2. What strategy helped you write the higher-level-cognitive-demand task? Explain.

3. In what ways will you support the implementation of the higher-level-cognitive-demand task during instruction?

Figure 1.8: Team discussion tool for identifying higher-level-cognitive-demand tasks for a unit.

Visit **go.solution-tree.com/mathematicsatwork** to download a reproducible version of this figure.

Before teachers on your collaborative teams use any higher-level-cognitive-demand task in class, each team should:

- Discuss its expectations for *student demonstration of quality work* in defense of the mathematical argument for the task (What will teachers expect to be a solution pathway?)

- Discuss how the lesson plan for the problem will be well-managed and promote communication of student argument with others and allow peer-to-peer–based solution defense

To help your collaborative teams facilitate this type of discussion, expect them to use figure 1.9 for any common higher-level-cognitive-demand task.

Preparing for the Use of Higher-Level-Cognitive-Demand Tasks

Once your teams include identifying and creating higher-level-cognitive-demand mathematical tasks as part of their unit planning, the before-the-unit activity will likely change. Your teachers will begin to look at mathematical tasks and problems, classify them as higher-level or lower-level cognitive demand, and then decide on the best points of entry during the unit for the higher-level-cognitive-demand tasks. The implementation and use of these tasks during the unit are discussed in further detail in chapter 2 in HLTA 6 (page 63).

Directions: Use these questions to better understand how you will use any higher-level-cognitive-demand task in class.
What is the essential standard for the lesson? (What do you want students to know and understand about mathematics as a result of this lesson?)
In what ways does the task build on students' previous knowledge? What definitions, concepts, or ideas do students need to know to begin to work on this task? What prompts will you need to help students access their prior knowledge?
What are all the possible solution pathways for the task?   Which of these pathways or strategies do you think students will use?   What misconceptions might students have?   What errors might students make?
What are the language demands of the task? How will you address these challenges if students are stuck during the task?
What are your expectations for students as they work on and complete this task? What tools or technology will they utilize to enhance student-to-student discourse?

Source: Adapted from Smith, Bill, & Hughes, 2008.

Figure 1.9: Task-analysis discussion tool.

Visit **go.solution-tree.com/mathematicsatwork** to download a reproducible version of this figure.

Team Progress

It is helpful to diagnose your collaborative teams' reality and action prior to launching the unit. You can use the before-the-unit status check tool for HLTA 2, identifying higher-level-cognitive-demand mathematical tasks, in table 1.2 as a diagnostic and feedback tool with your teacher teams. It matters less which stage your teams are at and more that team members are committed to working together to focus on understanding the learning standards and the best activities and strategies for increasing student understanding and achievement as teams seek stage IV—sustaining.

Of course, using balanced-cognitive-demand tasks becomes an important feature of the common assessment instruments for the end of the unit as well. Creating and using common assessment instruments with balanced-cognitive-demand tasks for each essential learning standard is the next high-leverage team action for your teacher teams' work.

Table 1.2: Before-the-Unit-Begins Status Check Tool for HLTA 2—Identifying Higher-Level-Cognitive-Demand Mathematical Tasks

Directions: Discuss your perception of your team's progress on the second high-leverage team action— identifying higher-level-cognitive-demand mathematical tasks. Defend your reasoning.			
Stage I: Pre-Initiating	**Stage II: Initiating**	**Stage III: Developing**	**Stage IV: Sustaining**
We do not discuss or share our use of the mathematical tasks in each unit of the curriculum.	We discuss and share some mathematical tasks we will use during the unit.	We explore and practice together mathematical tasks we will use during the unit.	We reach agreement on a collection of mathematical tasks every team member will use.
We do not share our understanding of the difference between lower- and higher-level-cognitive-demand mathematical tasks.	We do not base our instructional decisions and mathematical task choices on the cognitive demand of the task.	We are able to compare and contrast higher- and lower-level-cognitive-demand mathematical tasks for each learning standard of the unit.	We reach agreement on both the solution pathways for each mathematical task and the management of those tasks in the classroom.
We do not discuss the cognitive demand of the tasks we use in class.	We have reached agreement on what differentiates a higher- from a lower-level-cognitive-demand mathematical task.	We connect the mathematical tasks to the essential learning standards, daily lesson learning objectives, and corresponding activities for each unit.	We choose mathematical tasks that represent a balance of lower- and higher-level cognitive demand for the learning standards of the unit.
We do not use higher-level-cognitive-demand mathematical tasks.	We use higher-level-cognitive-demand mathematical tasks if they are included in the lesson.	We create higher-level-cognitive-demand mathematical tasks from lower-level-cognitive-demand mathematical tasks individually.	We create higher-level-cognitive-demand mathematical tasks from lower-level-cognitive-demand mathematical tasks as a team.

Visit **go.solution-tree.com/mathematicsatwork** to download a reproducible version of this table.

HLTA 3: Developing Common Assessment Instruments

One of the most powerful, high-leverage strategies for improving student learning is the creation of frequent, high-quality, common formative assessments.

—Richard DuFour, Rebecca DuFour, Robert Eaker, and Thomas Many

The mathematical tasks your teacher teams choose for their lessons partially answer the second critical question of a PLC—How will we know if they know it? The choices your teams make for the common assessment *instruments* they give students during and at the end of the unit fully answer this question.

As your teams make sense of the essential learning standards for the unit and better understand how to choose, adapt, and create higher-level-cognitive-demand mathematical tasks and learning activities, your teams will be ready to develop and use common assessment instruments to assess students' procedural fluency and understanding of the essential learning standards for the unit.

High-Leverage Team Action	1. What do we want all students to know and be able to do?	2. How will we know if they know it?	3. How will we respond if they don't know it?	4. How will we respond if they do know it?
Before-the-Unit Action				
HLTA 3. Developing common assessment instruments	▥	▥		

▥ = Fully addressed with high-leverage team action

▥ = Partially addressed with high-leverage team action

The What

────────── **Real-Life Leadership Scenario** ──────────

This high-leverage team action is a top priority of our Mathematics at Work team in every school we support. It is a first order of team business, so to speak. We were working in a more urban Midwest school district with K–5 teachers. The third-grade team was having a difficult time creating high-quality assessments, understandably so, after years of depending on the district office to do this for them. There was some push back and reluctance to do the work. Sometimes I [Tim] yield on an expected action—if I think the team may not be quite ready to take an important step like this in their growth. During this time, I was also speaking and working at Solution Tree's PLC Institutes. I was sitting on a panel with my colleague and friend Rick DuFour, who is the vision and voice of the PLC at Work framework. Rick made the comment to a member in the audience: "If you take away the real work teachers are to do in their teams, then you take away their learning." At that moment, I knew he was right. That third-grade team, the principal, and I were just going to have to dig in, learn, and grow, no matter how painfully, together. And we did—and so did the students.

Results in the school that spring soared as the teachers took greater ownership in the student assessments, the results of those assessments, and the instruction necessary to help students be prepared for learning the essential standards being assessed. And it all started by learning how to write high-quality common assessments together. The principal decided this should not be a math activity only for her teacher teams; they should begin to do the hard work of writing common assessments for everything taught in the school. It was my role to teach her the protocols for designing and evaluating high-quality mathematics assessments that should guide this aspect of teacher teamwork.

Why is developing common assessment instruments an important *before-the-unit* high-leverage team action for you to monitor? The process of creating *common* assessment instruments for each unit (and not just relying on the district assessments) for each grade level or course supports teacher conversations about prerequisite concepts and skills, common student errors, and ways of assessing students' understanding of the essential learning standards (including both content and process standards for the unit). It allows teachers to design lessons backward to connect the outcomes (student demonstrations of knowledge on essential learning standards) for the unit to the learning activities, tasks, and resources students use during the unit—what is needed for student success on the end-of-unit assessments.

According to DuFour, DuFour, Eaker, and Many (2010):

> One of the most powerful, high-leverage strategies for improving student learning available to schools is the creation of frequent, high-quality, common formative assessments by teachers who are working collaboratively to help a group of students acquire agreed-upon knowledge and skills. (p. 75)

In a PLC culture, this is a non-negotiable teacher team action.

However, there is an important distinction between formative assessment *processes* your teams use and the assessment *instruments* they use as part of that formative process. W. James Popham (2011) provides a powerful analogy to describe the difference between summative assessment instruments (such as end-of-unit tests) and formative assessment processes (such as what you and your students *do* with those test results). He describes the difference between a surfboard and surfing. The surfboard is a key component in the surfing process, but it is not the entire process. The entire process involves the surfer paddling out to an appropriate offshore location, selecting the right wave, choosing the most propitious moment to catch the chosen wave, standing upright on the surfboard, and staying upright while a curling wave rumbles toward shore.

Your teams' assessment instruments are some of the tools members use to collect data about student demonstrations of the essential learning standards. The assessment instruments subsequently will inform your teachers' and students' ongoing decisions about learning. Assessment instruments vary by grade level and can include such tools as teacher observations and interviews, exit slips, quizzes, unit tests, performance-item tasks, and in-class assignments.

However, to avoid inequity in the mathematics rigor for student learning, and to serve the formative learning process, these assessment instruments must be *in common* to each teacher on your grade-level or course-based team. When your collaborative teams create and adapt unit-by-unit common assessment instruments together, they enhance the coherence, focus, and fidelity to student learning expectations across all collaborative team members. (Team members will become less dependent on an outside authority such as the textbook or the district office.) The development of common assessments is so critical to student learning that it is specifically listed as one of the actions teachers should collaborate on under the professionalism principle in *Principles to Actions* (NCTM, 2014).

> You can help your teacher teams by monitoring the effective implementation of these common assessment instruments and to make sure they are used as intended, and not modified by individual teachers. You should collect them as they are designed, or redesigned, and review them for fidelity and quality. This is important as the assessments provide the hope for greater coherence, progression, and continuity for mathematics in the following year, and better student preparation.

Teachers minimize the wide variance in student task performance expectations (an inequity creator) from teacher to teacher by working collaboratively to design high-quality assessment instruments appropriate to the identified essential learning standards for the unit. Thus, the first questions you must ask each collaborative team are: "How do we know our end-of-unit assessments are of high quality? On what basis do we make these determinations?"

The How

Collaborative teams should consider the following when creating high-quality assessment instruments.

- What level of cognitive demand will we expect for each essential learning standard on the exam?

- What evidence of content knowledge will we assess for each essential learning standard?

- What evidence of student engagement in Mathematical Practices and processes will we assess for each essential learning standard?

- What types of question formats will we use to evaluate specific evidence of learning (such as multiple choice, short answer, multiple representations, explanation and justification, or the use of technology)?

Evaluating the Quality of Teachers' Current Assessment Instruments

One of the most important leadership actions you can take is to collect all of the unit-by-unit mathematics assessment instruments your teachers use. Are they of high quality, and how would you know? Figure 1.10 (page 34) is a quality-evaluation tool for during-the-unit or end-of-unit assessment instruments that your collaborative teams can use to evaluate the quality of the current unit assessment instruments, such as tests and quizzes, as well as to build new and revised assessment instruments for each unit of the course.

Assessment Indicators	Description of Level 1	Requirements of the Indicator Are Not Present	Limited Requirements of This Indicator Are Present	Substantially Meets the Requirements of the Indicator	Fully Achieves the Requirements of the Indicator	Description of Level 4
Identification and emphasis on essential learning standards (specific feedback to students)	Learning standards are unclear and absent from the assessment instrument. Too much attention is given to one target.	1	2	3	4	Learning standards are clear, included on the assessment, and connected to the assessment questions.
Visual presentation	Assessment instrument is sloppy, disorganized, difficult to read, and offers no room for work.	1	2	3	4	Assessment is neat, organized, easy to read, and well-spaced, with room for teacher feedback.
Balance of higher- and lower-level-cognitive-demand tasks	Emphasis is on procedural knowledge with minimal higher-level-cognitive-demand tasks for demonstration of understanding.	1	2	3	4	Test is rigor balanced with higher-level and lower-level-cognitive-demand tasks present.
Clarity of directions	Directions are missing and unclear. Directions are confusing for students.	1	2	3	4	Directions are appropriate and clear.
Variety of assessment task formats	Assessment contains only one type of questioning strategy, and no multiple choice or evidence of the Mathematical Practices. Calculator usage not clear.	1	2	3	4	Assessment includes a blend of assessment types and assesses Mathematical Practices modeling or use of tools. Calculator expectations are clear.
Tasks and vocabulary (attending to precision)	Wording is vague or misleading. Vocabulary and precision of language are a struggle for student understanding and access.	1	2	3	4	Vocabulary is direct, fair, accessible, and clearly understood by students, and they are expected to attend to precision in response.
Time allotment	Few students can complete the assessment in the time allowed.	1	2	3	4	Test can be successfully completed in the time allowed.
Appropriate scoring rubric (points)	Scoring rubric is not evident or is inappropriate for the assessment tasks presented.	1	2	3	4	Scoring rubric is clearly stated and appropriate for each task or problem.

Source: Adapted from Kanold, Kanold, & Larson, 2012, p. 94.

Figure 1.10: Assessment instrument quality-evaluation tool.

Visit **go.solution-tree.com/mathematicsatwork** to download a reproducible version of this figure.

A simple step would be to require each of your collaborative teams to rate and evaluate the quality of one of its most recent end-of-unit or chapter assessment instruments (tests). Using figure 1.10 and figure 1.11, the high-quality assessment diagnostic and discussion tool, ask members to evaluate the quality of the end-of-unit test as a team and report to you what they learned from the rating exercise.

How did it score? A 12? 16? 22? How close did the assessment instrument (the surfboard, so to speak) come to scoring a 27 or higher out of the 32 points possible in the rubric? You should expect that your teacher teams would write common assessment instruments that would score fours in all eight categories of the assessment evaluation rubric.

Directions: Examine your most recent end-of-unit assessment instrument, and evaluate its quality against the following eight criteria.

1. **Are the essential learning standards written on the test?** What do your students think about learning mathematics? Do your students think learning mathematics is about doing a bunch of problems? Or, can they explain the essential learning standards and perform on any task that might reflect that standard?

   Note: This is a necessary test feature if students are to respond to the end-of-unit assessment feedback when you pass it back. (See HLTA 9, page 103.)

2. **Does the visual presentation provide space for student work?** Do your students have plenty of space to write out solution pathways, show their work, and explain their thinking for each task on the assessment instrument?

   Note: This criterion often is one of the reasons not to use the written tests that come with your textbook series. You can use questions from the test bank aligned to your instruction, but space problems as needed.

3. **Is there a balance of higher- and lower-level-cognitive-demand questions on the test?** What percentage of assessment instrument tasks are lower-level cognitive demand? What percent are higher-level cognitive demand? Is there a balance? Unless this has been a major focus of your work, your current end-of-unit tests will not score very high in this criterion.

   Note: Underline the verbs on your test, and analyze what the verbs are asking the student to do. This will help you to better understand the level of cognitive demand. A good rule of thumb is that the rigor-balance ratio should be about 30/70 (higher- to lower-level cognitive demand).

4. **Is there clarity with all directions? What does clarity mean to each team member?** Are any of the directions for the tasks confusing to the student? Why?

   Note: The verbs used in your directions for each set of tasks are very important to notice when discussing clarity.

Figure 1.11: High-quality assessment diagnostic and discussion tool.   continued →

5. **Is there variety in the assessment formats?** Does your test use a blend of assessment formats or types? Did you include questions that allow for technology as a tool, such as graphing calculators? Did you balance the use of different question formats? If you use multiple choice, do you include items with multiple possible answers similar to those on the PARCC, SBAC, or other state assessments?

   Note: Your end-of-unit assessments should not be of either extreme—all multiple-choice or all open-ended questions.

6. **Is the language precise and accessible?** Is the vocabulary for each task in your end-of-unit assessment clear, accessible, and direct for students? Does it reflect the precision of language you use during the unit and that students understand?

   Note: Be sure the assessment instrument uses the proper language supports for all students.

7. **Is the allotted time appropriate?** Can your students complete this assessment in the time allowed? What will your procedure be if they cannot complete the assessment within the allotted time?

   Note: Each teacher on the team should complete a full solution key for the assessment. For upper-level students, it works well to use a time ratio of 3:1 (or 4:1) for student-to-teacher completion time to estimate how long it will take students to complete an assessment. (If it takes you ten minutes, you should allow thirty to forty minutes for the students to complete the assessment.) All teachers should use the agreed-on time allotment.

8. **Is the scoring rubric clear and appropriate?** Are the scoring rubrics for every task clearly stated on the test? Do your total points for the test make sense based on the tasks' complexity of reasoning? Are the tasks' points appropriate, and do all teachers agree on them?

   Note: See HLTA 4, page 41, for more details.

   Using your score from this assessment tool, which specific aspects of your current unit assessment instruments do you need to improve?

   _____

   _____

   _____

   _____

   _____

   _____

   _____

   _____

   _____

Figure 1.11: High-quality assessment diagnostic and discussion tool (continued).

Visit **go.solution-tree.com/mathematicsatwork** to download a reproducible version of this figure.

The value of any collaborative team–driven assessment depends on the extent to which the assessment instrument reflects the essential learning standards, can be used as part of a formative learning process in the aftermath of the assessment (see HLTA 9 in chapter 3 of this handbook), provides valid evidence of student learning, and results in a positive impact on student motivation and learning.

Designing a High-Quality Assessment Instrument

Providing time and expecting every teacher team to design common assessment instruments before the unit begins is a required aspect of your leadership. You will need to monitor this essential teacher team work as it provides a context for discussing prerequisite knowledge that teachers may need to address during instruction while simultaneously making sense of the learning standards and establishing a common expectation for student learning. It also provides a context for discussing where students might make common errors or hold common misconceptions. You will need to make sure teams have the necessary time, and then make this work a priority action (first order of business, so to speak) of your teams.

Your teams can use the questions in figure 1.12 (page 38) as a way to unpack an essential learning standard and prepare mathematical tasks and questions for teams' next common end-of-unit assessment instrument that they give to students.

Once teams identify and explore prerequisites and common errors, the teams are better prepared to find or develop the common assessment instrument tasks or questions. If this is a new activity for your collaborative teams, it might make sense to start with an existing assessment instrument and then ask teams to adapt it so that it addresses the essential learning standards comprehensively (listing them on the test itself for grades 3–12 and on the teacher's edition of the test for grades K–2) and provides an appropriate balance of higher- and lower-level-cognitive-demand tasks.

Perhaps the greatest challenge your teacher teams may face is balancing higher- and lower-level-cognitive-demand tasks on each unit exam. Cognitive-demand balance is the third criterion listed in the high-quality evaluation tool from figure 1.11 (page 35–36). When you review the end-of-unit assessment instruments your teams use (remember you are going to collect them from each grade- or course-level team for every unit), do the teams score well in terms of expected rigor balance? Did an appropriate blend of higher-level- and lower-level-cognitive-demand mathematical tasks and questions exist on the assessment?

Directions: Choose an essential learning standard you are planning to assess in your next end-of-unit assessment, and answer the following questions. Be sure to look very carefully at the verbs that describe the essential standard. They will provide hints about the question or task types you will need for the test.

Essential learning standard: _____

1. What prerequisite skills are necessary for this essential learning standard? How will you assess students' knowledge of these prerequisites?

2. What are common errors related to this essential learning standard? How will your instruction identify and resolve these errors before students take the common unit assessment?

3. How does your conversation around planning common assessment instruments influence your plans for instruction during the unit?

4. What mathematical tasks will you use during instruction for this essential standard, and what tasks will you reserve for the assessment of this standard?

Figure 1.12: Tool for planning and preparing for common assessment instrument task development.

Visit **go.solution-tree.com/mathematicsatwork** to download a reproducible version of this figure.

You can help your collaborative teams by asking them to use the questions from figure 1.13, checking for cognitive-demand balance on the common unit assessment instrument, as a general team discussion guide for creating an improved balance between higher- and lower-level-cognitive-demand tasks and student opportunities to identify common errors related to the essential learning standards.

Directions: With your collaborative team, answer the following questions to check the cognitive-demand balance of your common assessment instruments.

1. What does the current assessment instrument do well in terms of the nature of the cognitive demand for each mathematical task on the test?

2. How are prerequisite skills and common misconceptions regarding the essential learning standards addressed in this assessment instrument?

3. Which items on the assessment should remain as lower-level-cognitive-demand tasks?

4. Which items are more easily adapted into higher-level-cognitive-demand tasks? And how might you adapt them toward a higher level of cognitive demand?

Figure 1.13: Checking for cognitive-demand balance on the common unit assessment instrument.

Visit **go.solution-tree.com/mathematicsatwork** to download a reproducible version of this figure.

Your grade-level or course-based teams do not need to design unit assessments from scratch. You can use assessment instruments provided with your curriculum materials or the district and then discuss or adjust them to ensure they identify the essential learning standards, uncover common misconceptions, are appropriately balanced with respect to cognitive demand, provide the necessary time and space, and use appropriate, clear language and vocabulary.

Your collaborative teams could also create their own agreed-on criteria for assessment instrument quality using figure 1.11 (page 35–36) as a starting point based on your local vision for higher-level quality assessment. Use the tools in this chapter as a starting point for discussion, or adapt the tools if needed, to fit your vision for using high-quality assessment instruments in your mathematics program.

Just remember, your most important role is to collect copies of the tests teachers use, review them with key mathematics personnel in your school or district, give feedback about their quality, and make sure all teachers on the team are using them—that they are, in fact, *common.* You can visit **go.solution-tree .com/mathematicsatwork** to download sample exams appropriate to your grade level or course-based area of leadership.

Team Progress

It is helpful for your collaborative teams to diagnose their reality prior to launching the unit. Ask each team to assess its progress using the before-the-unit status check tool for HLTA 3, developing common assessment instruments, in table 1.3 (page 40) to determine how the collaborative team is currently functioning. Discuss your perception of your teams' progress on developing common assessment instruments. It matters less which stage your teams are at and more that team members are committed to working together to focus on understanding the learning standards and the best activities and strategies for increasing student understanding and achievement as your teams seek stage IV—sustaining.

Table 1.3: Before-the-Unit-Begins Status Check Tool for HLTA 3—Developing Common Assessment Instruments

Directions: Discuss your perception of your team's progress on the third high-leverage team action—developing common assessment instruments. Defend your reasoning.			
Stage I: Pre-Initiating	**Stage II: Initiating**	**Stage III: Developing**	**Stage IV: Sustaining**
We do not develop or use common assessment instruments.	Some members of our team develop common assessment instruments.	We develop common assessment instruments as a team, but not before the unit begins.	We design and write common assessments as a team before the unit begins.
We do not know if the end-of-unit assessments given by each member of the team are balanced for cognitive demand, provide sufficient time, and use clear language and vocabulary.	We develop end-of-unit common assessments connected to the learning standards, but they are not checked for balance of cognitive demand or clarity.	We develop common end-of-unit assessment instruments connected to the learning standards. They are either balanced for cognitive demand or clear but not both.	We develop common end-of-unit assessments that are clear, balanced, and connected to all aspects of the essential learning standards for the unit.
We do not know if our assessments are aligned to our instructional practices and reflect the essential learning standards of the unit.	We develop common assessments as a team, but not all members use them to influence their instructional plans for the unit.	Our planning for common assessments influences our instructional plans for the unit.	Our common assessments are deeply aligned with our instructional discussions and practices.

Visit **go.solution-tree.com/mathematicsatwork** to download a reproducible version of this table.

Once teams have prepared their common unit assessments, their efforts should turn to creating scoring rubrics for the test and developing expected proficiency expectations for students. Developing scoring rubrics and proficiency expectations for the common assessment instruments is the fourth high-leverage team action in the before-the-unit planning process. The process of developing scoring rubrics requires teams to reflect and stay focused on the essential learning standards for the unit. It also allows you to know whether or not there is a fidelity to the assessment of student work, no matter who the teacher for the grade level or course might be.

HLTA 4: Developing Scoring Rubrics and Proficiency Expectations for the Common Assessment Instruments

Do you trust me enough to allow me to grade your end-of-unit assessments?

—Timothy D. Kanold

How well do you monitor to determine whether those who teach the same course or grade level will score all assessment instruments with a *fidelity of accuracy*? Creating a team culture of collaborative assessment scoring is an important activity for your grade-level or course-based collaborative teams—and you must lead the creation of a culture that values this activity.

Just as the mathematical tasks and common assessment instruments (tests and quizzes) help your collaborative teams to answer the second critical question of a PLC—How will we know if they know it?—so do the choices your teams make for scoring the mathematical tasks on the common unit assessments. The purpose of this team action will be discussed further in chapter 3, HLTA 9 and 10 (see page 101).

High-Leverage Team Action	1. What do we want all students to know and be able to do?	2. How will we know if they know it?	3. How will we respond if they don't know it?	4. How will we respond if they do know it?
Before-the-Unit Action				
HLTA 4. Developing scoring rubrics and proficiency expectations for the common assessment instruments		◨		

◨ = Partially addressed with high-leverage team action

The What

──────────────── **Real-Life Leadership Scenario** ────────────────

In a recent school year, as the first grading period approached, a high school principal contacted me [Tim] and indicated she had received a number of complaints from concerned parents related to student grades in one section of advanced algebra because the grades were significantly lower than in other sections. There were three advanced algebra teachers in the building; I was puzzled because I knew all three to be experienced, effective teachers. In addition, I knew the team had agreed upon the essential learning standards (they were common), the curriculum included higher-level-cognitive-demand tasks, and that all three teachers used common formative assessments they had collaboratively developed.

In meeting with the advanced algebra team and discussing the results of their assessments, I asked to see the students' chapter tests from each of the teachers. What immediately became obvious was that one member of the team, the one with the lowest grades, was not grading her exams the same as the other two teachers.

The team thought they were implementing HTLA 4, but in reality they had only col-laboratively decided how many points to award each problem—they had not reached consensus on how to award those points. Two of the teachers, for example, were awarding students full credit for any student solution pathway that was mathemati-cally justified, while the third teacher was requiring students to solve problems in a specific manner and awarding points based on students' showing their work only for specific rote steps. This ultimately meant students in her class were not earning the same amount of credit as students in the other two classes for what were essentially the same solutions.

The team quickly realized that they had more work to do to ensure consistency in their grading practices and to help ensure that students' grades mean the same thing from class to class. The teacher who was not grading her tests the same way as the other two teachers at first refused to grade her tests like her two colleagues did, complaining that the "other teachers' standards were too low." I visited with the principal, who had to then work directly with the team for a period of time to ensure they all used common scoring rubrics. Eventually, as all members of the team under-stood the importance of common scoring rubrics, they began to develop and use them, grades between advanced algebra sections leveled out, and parent complaints about grading faded.

High-leverage team action 4 will improve the way your teachers provide instruction pathways for students during the unit. It will also improve the accuracy of their feedback and their grading practices at the end of the unit. Additionally, an extra benefit for you is the greater equity in the meaning and interpretation of student scores, regardless of the teacher for that grade level or course.

By expecting your teams to reach full agreement on the rubric score for each item on the end-of-unit test, you increase the reliability that the feedback for proficiency on the essential learning standards for the unit is accurate, and you increase the likelihood that all students understand the expectations of a solution pathway required to receive full credit on the task.

Most important, by monitoring this team action, you serve as an *inequity eraser* in your school or district, increasing the likelihood that teacher feedback on student performance will be consistent and accurate across all teachers in your school and that the same grades will mean the same level of learning has occurred from classroom to classroom in your building and district. This simple teacher team act has a very powerful student benefit for vertical learning from year to year and after eighth grade from course to course.

Determining how to score the mathematics tasks on an end-of-unit common assessment instrument involves far more than linking point values to test questions and tasks. As your teachers work together on scoring rubrics for the mathematical tasks, their instruction during the unit will benefit from:

- Discussing the value of each task relative to the other tasks on the test

- Deciding how they will determine if students have provided a complete solution for full credit relative to the essential learning standard each assessment task (problem) represents

- Deciding what they will do when students' answers are incomplete or incorrect—how will their response be scored?

These decisions are typically easier to make and more straightforward with lower-level-cognitive-demand tasks (consider your teachers' past mathematics unit assessment instruments) but not as clear for the higher-level-cognitive-demand tasks necessary to measure student understanding, reasoning, *and* procedural fluency.

The How

To help your teachers learn how to assign a score value to a mathematical task on a common assessment, ask them to examine a potential set of test questions (mathematical tasks), and engage in a collaborative discussion about the scoring of the task example.

If your collaborative teams do not have ongoing conversations about scoring, inequities will persist across the teams, and a grade (such as an A or B or a 1, 2, 3, or 4 proficiency) in one class will not be consistent with the grade in another class (another potential inequity). The examples that follow, around *one* test question only, emphasize that all teachers on the team need to discuss and practice *every* task (question) on the exam before giving the assessment instrument to students.

Your teams can use the collaborative team task-scoring discussion prompts in figure 1.14 (page 44) for any type of quiz or test tasks.

Ask your teams to practice using the prompts from figure 1.14 on the appropriate mathematical task in figure 1.15 (page 45). Most likely their answers will vary widely from one to three points, to four to six points, and to eight to ten points for full credit.

> As a school leader, your goal is not so much to be concerned with the exact number of points teachers assign to the mathematical task but rather that each team member is using the same scoring values. As students progress through the grade levels and experience higher-level-cognitive-demand mathematical tasks, collaborative team scoring becomes an even more pressing concern for your teacher teams.

Directions: With your collaborative team, discuss scoring for the task appropriate to your grade level from figure 1.15. Respond to each of the following questions.

1. What would you require students to demonstrate in order to receive full credit for the task? Think about solution pathways for the task.

2. How would you assign a score for proficiency to each part of a student's response?

3. How would you calibrate your individual responses in order to reach a scoring consensus agreement on the task with your collaborative team?

Figure 1.14: Collaborative team task-scoring discussion prompts.

Visit **go.solution-tree.com/mathematicsatwork** to download a reproducible version of this figure.

Grade 4 Task

Drag each expression to the correct column to show whether the product is less than or greater than 1.

Less Than 1	Greater Than 1

$3 \times \frac{1}{2}$	$5 \times \frac{1}{4}$	$1 \times \frac{1}{5}$	$4 \times \frac{3}{5}$	$2 \times \frac{2}{5}$

Grade 7 Algebra Task

Standard: I can understand the meaning of equivalent expressions.

Look at each expression. Is it equivalent to $\frac{x + 3y}{2}$?

Select Yes or No for expressions A–D.

A. $\dfrac{4x + 3y}{8}$ ○ Yes   ○ No

B. $\dfrac{5}{4}\left(\dfrac{2x + 6}{5}\right)$ ○ Yes   ○ No

C. $\dfrac{1}{2}(x + 3y)$ ○ Yes   ○ No

D. $\dfrac{2}{3}\left(\dfrac{5x}{6} + \dfrac{9y}{4} - \dfrac{x}{12}\right)$ ○ Yes   ○ No

Explain why each choice (A, B, C, and D) is equivalent or why it is not equivalent.

Grade II Algebra Task

A ball is thrown in the air. The height of the ball in terms of time is modeled by the graph shown. A second ball is thrown from a lower initial height and reaches a higher maximum height.

1. Select an equation that represents the height of the second ball in terms of time. Explain your reasoning.

$y = -x^2 + 2x + 3$	$y = x^2 - 3x + 4$	$y = -x^2 + 4x + 6$
$y = -x^2 + 5x + 3$	$y = x^2 - 4x + 2$	$y = -x^2 + 5x + 5$

2. What is the initial height of the second ball in terms of time?

3. What is the maximum height of the second ball?

Source: Smarter Balanced Assessment Consortium, 2013a; 2013b; 2013c. Used with permission.

Figure 1.15: Mathematical tasks for collaborative team task-scoring practice.

Creating a Team Scoring Rubric for Each Assessment

There is not one right answer when considering point values for the mathematics assessments given to students in your school. What is important is that you make sure your teams use the same scoring scale and base the scoring rubric on a decided standard (such as the complexity of reasoning the assessment task requires or a proficiency scale based on lower- or higher-level cognitive demand) and make sure that each collaborative grade-level or course-based team member honors the scoring scale. Part of your leadership work is to monitor this critical teacher team behavior. This equity pursuit will allow teams to hold the scoring of student work to a higher standard of accuracy, which we discuss in more detail with HLTA 9 in chapter 3 (page 103).

Your collaborative teams should use the prompts from figure 1.16 and answer the questions in order to calibrate their thinking regarding scoring a unit assessment. You can support this HLTA by asking your teams to occasionally provide their responses using the tool in figure 1.16 as you review their ongoing assessment work.

Directions: Within your collaborative team, answer each of the following questions in relation to each task on the assessment.

1. Which essential learning standard does each task address, and how do you know?

2. What do you expect students to demonstrate in order to successfully respond to and receive full credit for each task on the assessment?

3. How will you assign partial credit?

4. To which Mathematical Practices does each task develop? Describe why or why not.

5. What is the scoring value or point value assigned to each task, and how many total points should be used for this end-of-unit assessment?

6. Are there any questions on the test you would want to ask differently? If so, how would that impact the point value you would assign to the test question or task?

7. How many points correct would a student need for each essential learning standard in order to be considered proficient for that standard (the proficiency target)?

Figure 1.16: Assessment instrument alignment and scoring rubric tool.

Visit **go.solution-tree.com/mathematicsatwork** to download a reproducible version of this figure.

Setting Proficiency Targets

Notice the last question in figure 1.16: How many points correct would a student need for each essential learning standard in order to be considered proficient for that standard (the proficiency target)?

Your teams should decide what level of student performance will be considered proficient in each of the essential learning standards for the end-of-unit assessment. Your teams should know the learning score target they will expect each student to obtain for each learning standard to be considered proficient for that essential learning standard. Your teams' responses to students who do or do not achieve the learning proficiency target is the action described in HLTA 9 and discussed further in chapter 3 (page 101).

There is a type of standards-based grading practice gaining popularity in grades K–8 across the United States. It involves measuring students' proficiency on well-defined course learning standards (Marzano, 2010). Although many districts adopt standards-based grading *in addition* to traditional grades, standards-based grading can replace traditional point-based grades. If this is the case at your school, and you want more information on standards-based grading, you can go to www.marzanoresearch.com to review Robert Marzano's (2010) *Formative Assessment & Standards-Based Grading* or *A School Leader's Guide to Standards-Based Grading* (Heflebower, Hoegh, & Warrick, 2014) and learn more about the use of proficiency scales to score student work and measure student progress.

Team Progress

It is helpful to diagnose your collaborative teams' reality and action prior to launching the unit. Ask each team to assess its current reality using the before-the-unit status check tool for HLTA 4, developing scoring rubrics and proficiency expectations for the common assessment instruments, in table 1.4 (page 48). It matters less which stage your teams are at and more that team members are committed to working together and understanding the various student pathways for demonstrating solutions to the mathematical tasks on common assessments as your teams seek stage IV—sustaining.

These first four high-leverage team actions give your teams:

- A direct focus on your unit-by-unit understanding of and decisions regarding the essential learning standards
- Insight into the mathematical tasks and activities that support your work during the unit
- Understanding of common assessments to determine whether or not students have attained the content and process knowledge of the essential learning standards
- Guidelines for how to score student work and set proficiency expectations for each essential learning standard of the unit

There is one major high-leverage, equity-based team action left to complete before your teams launch into the unit and instruction: planning and using common homework assignments.

Table 1.4: Before-the-Unit-Begins Status Check Tool for HTLA 4—Developing Scoring Rubrics and Proficiency Expectations for the Common Assessment Instruments

Directions: Discuss your perception of your team's progress on the fourth high-leverage team action—developing scoring rubrics and proficiency expectations for the common assessment instruments. Defend your reasoning.

Stage I: Pre-Initiating	Stage II: Initiating	Stage III: Developing	Stage IV: Sustaining
We do not use common scoring rubrics on our assessments.	We discuss our scoring and grading practices collaboratively.	We create scoring rubrics for our common unit assessments collaboratively.	We create dependable scoring rubrics for all tasks on the common unit assessments as a collaborative team.
Each teacher establishes his or her own scoring system for their independent assessments.	We have not yet reached agreement on how to score the tasks on our common assessments.	We discuss and reach agreement on a student's complete response to receive full credit on each task for our common assessments.	We design assessment rubrics to align with students' reasoning about the mathematics for each essential learning standard of the unit.
We do not know the scoring and grading practices other members of our team use.	We use scoring rubrics independently, and do not discuss our use of scoring rubrics with other members of the team.	We use the common end-of-unit assessment scoring rubrics for measuring student proficiency on each learning standard but don't discuss them as a team.	We use the common end-of-unit assessment scoring rubrics for measuring student proficiency on each learning standard and discuss them as a team.
We do not set student proficiency targets for each essential learning standard of the unit.	We set student proficiency targets independently, but do not know the proficiency targets other members of our team use for each essential learning standard of the unit.	We collaboratively set student proficiency target performances on the end-of-unit assessment for some, but not all, of the essential learning standards of the unit.	We collaboratively set student proficiency target performances on the end-of-unit assessment for each essential learning standard of the unit.

Visit **go.solution-tree.com/mathematicsatwork** to download a reproducible version of this table.

HLTA 5: Planning and Using Common Homework Assignments

Assign work that is worthy of their best effort (problem solving and reasoning).

—Linda Darling-Hammond

By using homework for practice in self-assessment and complex thinking skills, we can put students in charge of the learning process.

—Cathy Vatterott

The mathematical tasks and problems your teachers assign *as homework* should help students to accurately answer the question, "How will I know if I understand the essential learning standards of the unit each and every day?" Thus, you need to ensure your teacher teams reach agreement on the purpose, coherence, rigor, and length of time for homework. In addition, teams need to agree on how to use common homework assignments and communicate this to students, parents, and support staff. Once again, your teams' work to develop common homework assignments for the unit, before it begins, can help eliminate another potential inequity creator that you can monitor for and help to erase.

You should plan on collecting evidence of the homework assignments given to students for every unit. This is another way your teams reach agreement on the second critical question of a PLC: How will we know if they know it?

High-Leverage Team Action	1. What do we want all students to know and be able to do?	2. How will we know if they know it?	3. How will we respond if they don't know it?	4. How will we respond if they do know it?
Before-the-Unit Action				
HLTA 5. Planning and using common homework assignments	▣	▣	▣	▣

▢ = Fully addressed with high-leverage team action

▢ = Partially addressed with high-leverage team action

The What

––––––––––––––––––––– **Real-Life Leadership Scenario** –––––––––––––––––––––

One year at the annual NCTM conference, I [Tim] presented a session on the topic of mathematics homework. At the end of the session, there was a long line of teachers with questions, as this can be a very difficult topic to address. Two fourth-grade teachers from Sweden came up to me and said, "You butchered the homework!" I was not sure if this was good or bad. I took a deep breath and asked what they meant. They explained, "You took the knife and stabbed the beast!" At that moment, I understood the need to tackle the topic—to eliminate the secrecy around how teachers decide to assign independent practice to students and to make the creation and assignment of mathematics homework a collaborative team effort for the grade level or course. The young men from Sweden understood. I left wondering how many school principals actually monitor and know the exact mathematics homework (quality and quantity) teachers assign to students in their building each and every day. Why are the decisions about mathematics homework (independent and measured student practice) in most schools such a private teacher act?

Planning and using common homework assignments is an important before-the-unit high-leverage team activity because mathematics homework is often an instructional area that lacks clarity, purpose, and certainty for students, parents, intervention support personnel, and most important, for you as a school leader. You should help your teams by asking them, "Why do we give students mathematics homework? What is the purpose of mathematics homework? Why won't students do their mathematics homework? Why do you spend so much time going over mathematics homework in class? Why is mathematics homework assigned for a grade?"

The very idea of homework, and what to do with it, is often a conundrum for you, your teachers, and your students. Is homework really an essential element to the process of student learning? The short answer is yes, but the best protocols to follow for homework are not quite as clear.

However, four things are clear.

1. The assignment of independent practice or homework cannot be a superficial exercise for your teacher teams or your students.

2. Anyone who is an expert at anything devotes significant time to practice (Gladwell, 2008).

3. If teachers deny students an opportunity for independent practice, they deny them the very thing they need to develop real competence (Anderson, Reder, & Simon, 1995).

4. Homework assignments need to be spaced and in common.

The homework teachers assign to students as well as the way they use homework as an in-class activity, in other words, as a formative task to guide instruction, needs to be a carefully thought-out and planned-for *team* discussion, agreement, and activity *before* the unit begins (Gladwell, 2008).

Although research on homework does not support a clear path for all grades and all subjects (Cooper, 2008a, 2008b), the issue of homework becomes more complicated as attention turns to implementing the Common Core State Standards and other more rigorous state standards and understanding them (and thus higher-level-cognitive-demand practice problems).

Research does indicate that homework can be helpful in improving student achievement if implemented appropriately (Cooper, 2008b). A key finding from the research is that homework is most effective when teachers provide feedback to students' homework on a daily basis and give students written descriptive feedback that goes beyond simply marking student work as correct or incorrect (An, 2004; Davies, 2007; Marzano, 2006).

Practice is important but not without first developing student understanding. Practice without understanding may be detrimental to students' development of fluency, and in many cases, avoiding this danger means that instruction should place greater emphasis on guided practice—practice that is supported by monitoring and feedback—prior to independent practice (Larson, 2011). Marzano (2007) finds that to have a positive effect, homework should also have a clear *purpose* that teachers communicate to students: to deepen students' conceptual understanding, enhance procedural fluencies, or to allow students an opportunity for independent formative practice around higher-level-cognitive-demand tasks. Teachers should intentionally consider and choose each homework problem or mathematical task carefully based on the essential learning standards of the lesson and unit.

Research also supports the idea of *spaced* (sometimes called *distributed* or *spiral*) versus *massed* homework practice during the unit of study (Hattie, 2012; Pashler, Rohrer, Cepeda, & Carpenter, 2007) as having a significant impact on student learning. That is, teachers should provide homework assignment (practice) tasks that are spaced throughout the unit, allowing students to cycle back and perform distributed practice on prior learning standards, including those learned earlier in the unit, in previous units, or possibly in the previous grade level.

Teams should design practice for student learning *outside of class* (homework) from the same core set of problems the team agreed on, thereby ensuring that each student, no matter what teacher they have for the course, would work to meet the same expectations.

The How

Collaborative team discussions regarding the role of homework and the selection of homework problems can be a powerful professional growth experience and should be an embedded part of your teacher teams' work every unit.

> In general, you should expect your team leaders to provide you, your students, and parents with complete homework assignment sets before each unit begins, and you should monitor to ensure the mathematics homework in your school or district meets the standards described in this section of the handbook.

Understanding the Purpose of Homework

You will need to help your teachers to understand the purpose of daily mathematics homework. You can use the collaborative homework assignment protocol discussion tool in figure 1.17 (page 52) to help your teams develop a better understanding of the purpose, content, and expected protocols for the unit's homework assignments. These prompts can also be used for team discussion with vertical course-based teams as you examine mathematics homework protocols and progressions across all grade levels or courses in your building.

Directions: Use the following prompts to guide discussion of the unit's homework assignments.

Purpose of homework:

1. Why do we assign homework for each unit's lessons? What is the purpose of homework?

Nature of homework:

2. What is the proper number of mathematical tasks for daily homework assigned during the unit? In other words, how much time should students spend on homework?

3. What is the proper rigor (cognitive-demand expectations) of the mathematical tasks for homework assigned during the unit?

4. What is the proper distribution of tasks for homework to ensure spaced practice (cyclical review) for our students?

5. How do our daily homework assignments align to the learning standard expectations for the unit?

6. How will we reach consensus on unit homework assignments in order to ensure coherence to the student learning and practice expectations?

Use of homework:

7. How should we grade or score homework assignments?

8. What will we do if students do not complete their homework assignments?

9. How will we go over the homework in class?

10. How will we communicate the common unit homework assignments to students, parents, and support staff?

Figure 1.17: Collaborative homework assignment protocol discussion tool.

Visit **go.solution-tree.com/mathematicsatwork** to download a reproducible version of this figure.

It is the expectation that your collaborative teams will reach full agreement on their responses in figure 1.17 before the unit begins to select appropriate independent practice task (homework problem) sets for students. You should collect and monitor these homework sets.

In response to figure 1.17, question one, Why do we assign homework for each unit's lessons? What is the purpose of homework?, it is important to note that the primary purpose of homework is *not summative*. Homework should not be assigned to students in order to assign a grade. In fact, homework should generally not count for more than 5 to 10 percent of the total student grade. Because homework is a *formative* activity—an opportunity to obtain feedback and improve learning—it should not constitute so much of a student's grade that it is not reflective of actual class performance at the end of a learning cycle.

The primary purpose of mathematics homework is *independent student practice*. More importantly, *successful* independent practice. That is, students must understand and use homework as an opportunity for a self-guided formative feedback learning process (Hattie, 2012). Independent practice can be with other students, it can be with other adults, it can be with help from YouTube or other social media resources. However, while outside of class, away from their teacher and guided in-class practice, students must practice mathematics problems and connect those problems to the essential learning standards.

Students need to come to understand that they complete homework, *not* because they will receive a grade, *not* because they hope the teacher will go over the problems in class the next day (which makes homework no longer an independent practice exercise), *not* because they are being punished, but only because they understand the importance of formative assessment and successful practice as a critical part of their long-term learning process. In class, students do need teacher modeling and lots of peer-to-peer guided practice (see HLTA 7 in chapter 2 of this handbook, page 75), and then outside of class, and in a timely fashion, students need to perform accurate independent practice with feedback (self-feedback or with peers) and then take action—well before they are back in class the next day.

Your role is to provide your collaborative teams with the time needed to determine how much homework to provide for additional student independent practice. Your teams must decide how to communicate the homework assignments to students and parents and the role homework plays as part of the classroom protocols described in figure 1.18 (page 54).

Again, this all should occur *before* the new unit of study begins. How will you know? Perhaps the homework paradigm shift for you and your teachers is to stop calling it *homework*. Certainly *independent practice* can be done at the coffee shop, after school in a classroom, on the bus, sitting in a hallway, in the car on the way to practice or a game, with friends at the library, or with parents; it does not have to be done at home.

Using Effective Homework Protocols

From a rigor, coherence, and equity point of view, the homework your teacher teams assign for a unit of study needs to be the same for all students in each mathematics course in your building. Teachers should give team-developed homework assignments to students, parents, and support staff in advance of teaching the unit with the understanding that teams can and will modify the assignments during the unit as necessary to address specific student learning needs. To check the temperature on how your teams are currently responding to the homework dilemma, you can use the homework quality diagnostic tool in figure 1.18 to measure team progress.

High-Quality Homework Indicators	Description of Level 1	Requirements of the Indicator Are Not Present	Limited Requirements of This Indicator Are Present	Substantially Meets the Requirements of the Indicator	Fully Achieves the Requirements of the Indicator	Description of Level 4
The primary purpose of homework is independent practice.	Homework is primarily assigned to give a student a grade. Homework counts more than 10 percent of a student's total grade.	1	2	3	4	Homework is understood as primarily for independent practice and a formative assessment learning loop for students. Homework counts no more than 10 percent of a student's grade.
Homework assignments are the same for every teacher on the course team.	Each teacher on the team creates his or her own homework assignments and does not share with others.	1	2	3	4	Common homework assignments are developed collaboratively by the team and are the same for all students in the grade level or course.
All homework assignments for the unit are given to the students before the unit begins.	Students find out homework assignments each day or each week as the unit progresses.	1	2	3	4	Students are provided all unit homework assignments–electronically or with a handout—as the unit begins.
Homework assignments for the unit are appropriately balanced for cognitive demand.	Homework practice problems are not balanced for rigor. Emphasis is on lower-cognitive-demand tasks.	1	2	3	4	Homework practice is appropriately balanced with higher- and lower-cognitive-demand tasks.
All practice problem answers are given to the students in advance of the homework assignments.	Students must wait until the next day to receive answers or solutions to homework practice problems.	1	2	3	4	Students are able to check their solutions during independent practice and are expected to rework the problems if not correct the first time.
Homework assignments for each unit exhibit spaced and massed practice.	The homework assignments represent superficial thought as to the problems chosen and consist of massed practice.	1	2	3	4	The homework assignments represent carefully chosen problems or tasks. Spaced practice from several lessons of the unit or previous units is included in addition to massed practice.
Daily homework is aligned to the essential learning standards of the unit.	Students are not able to make connections between the daily homework practice problems and the learning standards of the unit.	1	2	3	4	Students connect the homework practice as essential to helping them demonstrate knowledge of the essential learning standards of the unit.
Limited time is spent going over homework in class.	Students and the teacher spend fifteen to twenty-five minutes (or more) in class going over the homework answers and solutions. The teacher does most of the work as the students watch.	1	2	3	4	At most, five to seven minutes of class time are used discussing the homework. It is primarily a peer-to-peer class activity facilitated by the teacher.

Figure 1.18: Homework quality diagnostic tool.

Visit **go.solution-tree.com/mathematicsatwork** to download a reproducible version of this figure.

Ask your teams to use the diagnostic tool from figure 1.18 to give a score to the quality of their homework assignments for a given unit. The maximum score is 32. Note that teams should list the essential learning standards for the unit on the homework assignment sheet. Note too that the problems assigned for each lesson should illustrate spaced practice and not massed practice during the unit. You should monitor for evidence of these criteria.

Also, determine whether or not students (if age appropriate) or their parents have access to the mathematics homework assignment set online for each grade level or course as a Google Doc or some other web source. Finally, be sure to limit the amount of time (most days no more than five minutes of class time) spent going over the homework in class.

To summarize, you can guide collaborative team homework protocol development by asking your teacher teams to consider the following benchmarks.

- **Homework purpose:** The primary purpose of homework should be to allow the student the opportunity for *independent practice* on learning standards mastered in class during guided practice and small-group discourse. Homework can also provide a chance for the student to practice mathematical tasks that relate to previous learning standards or tasks that reflect prerequisite learning standards for the next unit. Homework that provides review of previous work and helps to prepare students for future work leads to improved student achievement (Cooper, 2008a).

- **Homework length:** How much time should daily homework take students to complete? How many problems should it entail? Homework should not be lengthy (Cooper, 2008b), so teachers should take care about what they assign. Take into account the cognitive demand of the tasks or problems you assign. Homework tasks as a general rule should not take more than thirty to thirty-five minutes (per course) of time outside of class.

- **Homework task selection:** The homework your school curriculum or textbook includes is not necessarily appropriate for students without some adjustments with which each team agrees. Make sure that all tasks are necessary as part of independent practice, have spaced practice and not massed practice, and align to the stated learning goals of the unit.

- **Homework answers:** There are many advantages to providing students with homework answers before the unit of instruction begins. When teachers provide students with answers to the homework problems, they can check their solutions against the answers, and if their end results do not match the provided answers, they can rework the problem to find their errors. In other words, students receive immediate and formative self-assessed feedback of their work—like when playing an electronic game. Moreover, a compelling reason to provide students with the answers to the homework in advance of the assignment is to save time during the class period the next day. No time should be spent going over the answers or the actual homework problems. Remember, homework is *independent* practice, not *in-class* practice. Since the students know exactly what they know and what they do not understand, any in-class discussion time on homework can be limited to a brief few minutes and becomes more meaningful for the students.

continued →

> • **Homework focus in class:** Once each collaborative team determines homework, teachers can focus on how to address homework in class, the type of feedback that they will give students, and what will occur if students do not complete the homework. If teachers spend most of the class time going over homework, you lose the impact of successful independent practice on student learning. Students may be choosing to wait to do homework problems because they know they can write down the work when you go over the problems the next day. Since the purpose of homework is independent practice, limit the amount of time in class to grade, score, or go over the practice problems. If teachers spend most of the class time going over homework, each team must revisit the amount and content of what you assign. It could be that the team assigned too much homework or that students did not achieve an appropriate level of mastery prior to practice of the learning standard.

These homework processes and procedures should be consistent from teacher to teacher within each of your collaborative teams. Taking the time to monitor the quality of the homework assigned and how it is used in class is an important assessment issue and part of your leadership role in the school or district.

Team Progress

It is helpful to diagnose your teams' current reality and provide feedback to help them take corrective action prior to launching the unit. Ask each team to individually assess its progress using the status check tool in table 1.5. It matters less what stage teams are at and more that members are committed to collaboratively defining the purpose of homework, using the same common homework assignments and protocols, and communicating those assignments to students, parents, and colleagues as your teams seek stage IV—sustaining.

Table 1.5: Before-the-Unit-Begins Status Check Tool for HLTA 5—Planning and Using Common Homework Assignments

Directions: Discuss your perception of your team's progress on the fifth high-leverage team action—planning and using common homework assignments. Defend your reasoning.			
Stage I: Pre-Initiating	**Stage II: Initiating**	**Stage III: Developing**	**Stage IV: Sustaining**
We do not have a clear purpose for why we assign homework.	We have *established* a clear purpose for homework, but it is not independent and formative student practice.	We have *developed* the shared purpose of using homework as independent formative student practice.	We have *implemented* the shared purpose of homework as independent formative student practice.
We do not plan or use common homework assignments and do not know the homework assignments given by other members of our team.	We discuss homework assignments and have not yet reached collaborative agreement on the nature of those assignments for each unit.	We collaboratively *plan* and develop common homework assignments for each unit.	We collaboratively *use* common homework assignments for each unit.
We do not know the nature of the homework protocols used for the assignments given by other members of our team.	We discuss the nature of the homework protocols used for the assignments given by other members of our team, but do not agree on those protocols.	We have team agreement on developed homework protocols including limited number of tasks, spaced practice, balance of cognitive demand, and alignment to the essential learning standards.	We have complete team agreement on homework protocols including limited number of tasks, spaced practice, balance of cognitive demand, and alignment to the essential learning standards, and we use those protocols with our students.
We do not know how other members of our team go over homework in class.	We discuss how we go over homework in class but do not agree on what we should do.	We discuss how we go over homework in class and agree on what we should do with homework during class.	We discuss how we go over homework in class, agree on what we should do, and implement that agreement.
We do not know how other members of our team count homework as a percent of the student's total grade.	We know how others count homework for a grade, but we each do it our own way.	We grade homework the same each day, but we count it differently from other team members as a percent of the total student grade.	We have complete team agreement on how homework should be used and accounted for as part of the student's total grade.

Visit **go.solution-tree.com/mathematicsatwork** to download a reproducible version of this table.

Setting Your Before-the-Unit Priorities for Team Action

When your school functions within a PLC culture, your grade- and course-level teams make the commitment to reach agreement on the five high-leverage team actions outlined in this chapter, and you have helped them in this endeavor.

> HLTA 1.   Making sense of the agreed-on essential learning standards (content and practices) and pacing
>
> HLTA 2.   Identifying higher-level-cognitive-demand mathematical tasks
>
> HLTA 3.   Developing common assessment instruments
>
> HLTA 4.   Developing scoring rubrics and proficiency expectations for the common assessment instruments
>
> HLTA 5.   Planning and using common homework assignments

Your teams cannot focus on everything and will not have the time to do so for every unit. You should help your teams set priorities. You can use figure 1.19 to focus time and energy on actions that are most urgent in your teams' preparation for each unit during the year. Less is more. Focus on fewer things, and make those things matter at a deep level of team implementation.

These five high-leverage team actions combine to form step one of the teaching-assessing-learning cycle (see figure 1.1, page 10) and will help you prepare teams for the rigors and challenges of teaching and learning during the unit. They are also linked to teacher actions that will significantly impact student learning in each class.

In chapter 2, we turn our attention to steps two and three of the teaching-assessing-learning cycle and the mathematics instruction that takes place in your school or district. Here we focus on implementing in-class formative assessment processes through teaching strategies that include the use of higher-level-cognitive-demand mathematical tasks, common end-of-unit assessment instruments, and homework that is strongly connected and aligned to the learning standards for the unit.

Directions: Identify the stage you rated your team for each of the five high-leverage team actions, and provide a brief rationale. When you are ready, discuss your ratings as a team.

1. Making sense of the agreed-on essential learning standards (content and practices) and pacing

   Stage I: Pre-Initiating   Stage II: Initiating   Stage III: Developing   Stage IV: Sustaining

   Reason: _____

   _____

   _____

2. Identifying higher-level-cognitive-demand mathematical tasks

   Stage I: Pre-Initiating   Stage II: Initiating   Stage III: Developing   Stage IV: Sustaining

   Reason: _____

   _____

   _____

3. Developing common assessment instruments

   Stage I: Pre-Initiating   Stage II: Initiating   Stage III: Developing   Stage IV: Sustaining

   Reason: _____

   _____

   _____

4. Developing scoring rubrics and proficiency expectations for the common assessment instruments

   Stage I: Pre-Initiating   Stage II: Initiating   Stage III: Developing   Stage IV: Sustaining

   Reason: _____

   _____

   _____

5. Planning and using common homework assignments

   Stage I: Pre-Initiating   Stage II: Initiating   Stage III: Developing   Stage IV: Sustaining

   Reason: _____

   _____

   _____

With your collaborative team, respond to the red light, yellow light, and green light prompts for the high-leverage team actions that you and your team believe are most urgent.

Red light: Indicate one activity you will stop doing that limits effective implementation of each high-leverage team action.

Yellow light: Indicate one activity you will continue to do to be effective with each high-leverage team action.

Green light: Indicate one activity you will begin to do immediately to become more effective with each high-leverage team action.

Figure 1.19: Setting your collaborative team's before-the-unit priorities.

Visit **go.solution-tree.com/mathematicsatwork** to download a reproducible version of this figure.

CHAPTER 2

During the Unit

The choice of classroom instruction and learning activities to maximize the outcome of surface knowledge and deeper processes is a hallmark of quality teaching.

—Mary Kennedy

Learning is experience. Everything else is just information.

—Albert Einstein

Much of the daily work of your teacher teams occurs during the unit of instruction. This makes sense, as it is during the unit that teachers place into action much of the team effort put forth in their before-the-unit work.

Your role is to support teachers' efforts in data gathering, sharing, feedback, and action regarding student learning that forms the basis of an in-class *formative assessment process* throughout the unit. The teacher sharing of in-class formative assessment processes provides the platform that allows your collaborative teams to make needed adjustments to instruction, tasks, and activities that will better support student learning during the unit.

A teacher team–led effective formative assessment process also empowers and motivates students to make needed adjustments during class in their ways of thinking about and doing mathematics that lead to further learning.

This chapter is designed to help you support your collaborative teams' members prepare and organize discussions and lessons around three additional high-leverage team actions. These three high-leverage team actions are based on steps two and three of the PLC teaching-assessing-learning cycle in figure 2.1 (page 62).

The three high-leverage team actions that occur during the unit of instruction are the following.

> HLTA 6. Using higher-level-cognitive-demand mathematical tasks effectively
>
> HLTA 7. Using in-class formative assessment processes effectively
>
> HLTA 8. Using a lesson-design process for lesson planning and collective team inquiry

Steps two and three of the teaching-assessing-learning cycle provide a focus for your teachers' use of effective in-class lesson strategies. How well your collaborative teams implement formative feedback and assessment processes (step two) is only as effective as how well each teacher is able to elicit student actions and responses to the formative feedback teachers and student peers provide to one another (step three). This type of instructional vision is the most important form of feedback you can give your teachers during classroom walkthroughs and observations of teacher practice.

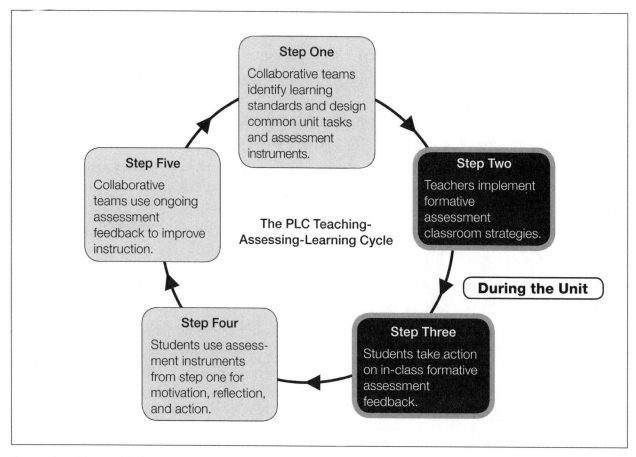

Source: Kanold, Kanold, & Larson, 2012.

Figure 2.1: Steps two and three of the PLC teaching-assessing-learning cycle.

HLTA 6: Using Higher-Level-Cognitive-Demand Mathematical Tasks Effectively

This sixth high-leverage-team action highlights the teacher teams' work to present, adjust, and use daily common higher- and lower-level-cognitive-demand mathematical tasks. Teams designed these tasks in step one of the teaching-assessing-learning cycle as part of HLTA 2 (page 20). Now, teachers apply them in an engaging and formative student learning environment.

Recall there are four critical questions every collaborative team in a PLC culture asks and answers on an ongoing unit-by-unit basis.

1. What do we want all students to know and be able to do? (The essential learning standards)

2. How will we know if they know it? (The assessment instruments and tasks teams use)

3. How will we respond if they don't know it? (Formative assessment processes for intervention)

4. How will we respond if they do know it? (Formative assessment processes for extension and enrichment)

The sixth HLTA—using higher-level-cognitive-demand mathematical tasks effectively—ensures your teams reach clarity on the second PLC critical question, How will we know if they know it?

High-Leverage Team Action	1. What do we want all students to know and be able to do?	2. How will we know if they know it?	3. How will we respond if they don't know it?	4. How will we respond if they do know it?
During-the-Unit Action				
HLTA 6. Using higher-level-cognitive-demand mathematical tasks effectively	▨▢	▨		

▨ = Fully addressed with high-leverage team action

▢ = Partially addressed with high-leverage team action

As a school leader, you might find that this high-leverage team action is a bit more difficult to monitor. You need to make sure every grade-level or course-based team in your area of leadership intentionally plans for, designs, and implements mathematical tasks that will provide student engagement and descriptive feedback around key elements of the learning standards and Mathematical Practices for the unit. Yet mathematics may not be part of your content background. Regardless, the tools in this section can help you monitor for this important team action.

The What

--------------------------- **Real-Life Leadership Scenario** ---------------------------

I [Tim] was working with a high school algebra 2 team in Minnesota—a great team of teachers. The teachers cared about students and about student learning. They were caught up, however, in that strange transition phase where the verbs in their standards were shifting to the word understand. *For example, "Students will identify the end behavior of a function" was the old standard versus "Students will* understand *the end behavior of a function" as the new standard.*

After several classroom visits to each teacher on the team, I was struck by the low-level tasks they asked students to perform each day in class and students' lack of productive struggle. Students in their classes either worked hard to complete a lower-level mathematics task or shut down and waited for the teacher to show them what to do.

At a team meeting, I asked the teachers why they thought this was happening. Their response was interesting: they hadn't really thought much about it. Although the standard itself had changed intent (demonstrate understanding), they thought that if the students could do the lower-level-cognitive-demand mathematics problems, they must understand. If students were stuck, their primary teacher response should be to show the students how *to do it. (*How *speaks to procedure.) It had not occurred to them that they might be asking the students to do the wrong set of mathematical tasks or that the desire to immediately show the students how to do the mathematics tasks was causing students to avoid productive struggle.*

The team was also struggling to decide how much to expect students to attend to precision—an expectation of the process standards. We had a long discussion about whether or not it was okay for students to use phrases like, "It goes up or it goes down" to describe the end behavior of a function. This would become more of an issue as teachers on the team began to use mathematical tasks that required students to justify their reasoning. I talked to the department chair and explained that his team needed to spend more team time talking about how they use tasks in class (see the protocols that follow in this HLTA).

When students work on cognitively demanding tasks, they often struggle at first. Some mathematics teachers and school administrators observing those teachers perceive student struggle as an indicator that the teacher has failed instructionally. Thus, the teacher will jump in to rescue students by breaking down the task and guiding students step by step to a solution (NCTM, 2014). This, in turn, deprives students of an opportunity to make sense of the mathematics (Stein et al., 2007) and does not support student engagement with the Standards for Mathematical Practice, such as with Mathematical Practice 1: "Make sense of problems and persevere in solving them."

As a result of the collaborative teacher team *work* on high-leverage team action 6, students will begin to see that productive struggle, a research informed instructional strategy, is an important part of learning mathematics (NCTM, 2014), and teachers will begin to develop student proficiency in the Standards for Mathematical Practice and processes by using higher-level-cognitive-demand tasks in class that support learning the essential standards of the unit.

You will need to monitor the during-the-unit teacher team discussions. Do your teachers begin to see the value of productive struggle as an important part of learning mathematics? You should support and expect to see students engaged in productive struggle during the daily use of higher-level-cognitive-demand tasks in class that support learning the essential standards of the unit.

HLTA 6 consists of three action components.

1. Observing student proficiency in the Mathematical Practices and processes
2. Observing the effective use of higher-level-cognitive-demand tasks during the unit of instruction
3. Observing a sustained focus on the essential learning standards during the unit of instruction

This high-leverage team action begins by revisiting and exploring what it means for students to be proficient in the Standards for Mathematical Practice—the practices that describe *how* students should engage with the mathematics during class. Interestingly, how students engage in and experience learning mathematics was a critical expectation of mathematics instruction long before the CCSS came along. The process of learning mathematics, and not just the content standards for learning, has been a part of the guaranteed and viable curriculum since NCTM's (1991) *Professional Standards for Teaching Mathematics*. The CCSS merely serve as a catalyst that places a more direct focus onto the conversation.

To review, the CCSS for mathematics provide eight Standards for Mathematical Practice. The eight Standards for Mathematical Practice (presented in more detail in appendices A and B, pages 131–138) are (NGA & CCSSO, 2010, pp. 6–8):

1. Make sense of problems and persevere in solving them.
2. Reason abstractly and quantitatively.
3. Construct viable arguments and critique the reasoning of others.
4. Model with mathematics.
5. Use appropriate tools strategically.
6. Attend to precision.
7. Look for and make use of structure.
8. Look for and express regularity in repeated reasoning.

The How

In chapter 2 of *Common Core Mathematics in a PLC at Work, Leader's Guide* (Kanold, Kanold, & Larson, 2012), we provide three key questions to help your teams better understand the Standards for Mathematical Practice.

1. What is the intent of the Mathematical Practice, and why is it important?
2. What teacher actions facilitate student engagement in this Mathematical Practice?
3. What evidence exists that students are demonstrating this Mathematical Practice?

It should be your expectation that *every teacher* can eventually respond accurately and intelligently to these three questions.

Observing Student Proficiency in the Mathematical Practices and Processes

It is important to note that effective leadership of deep teacher understanding of the Mathematical Practices involves observing how students *learn and do* mathematics. Whether or not your state adopted the Common Core, is a member of one of the Common Core assessment consortia, or has established its own state standards and assessments is not essential to this high-leverage team action. Research about how students learn mathematics at high levels of achievement is the basis on which the Mathematical Practices and processes stand (Hattie, 2012; Kilpatrick, Swafford, & Findell, 2001; Martin, 2007). In short, deep teacher understanding of how to develop student proficiency in these practices has a significant learning benefit to the student, regardless of the nature of your state standards or testing criteria. (See appendix D, page 141, or visit **go.solution-tree.com/mathematicsatwork** to access research resources related to the ten high-leverage team actions in this handbook.)

Your faculty will need to develop consensus on the meaning of *proficiency* relative to student engagement with the Mathematical Practices or processes (used by your state) during the unit. To support this work, if you are from a state that addresses the Common Core State Standards, ask your teams to identify proficiency with the Standards for Mathematical Practice using the tool in figure 2.2. This tool is a way of organizing the initial ideas teams generate around the meaning of proficiency for each of the eight Mathematical Practices. For an in-depth response to the types of evidence that your teams might have brainstormed, see appendix B (page 135). You can compare the ideas in appendix B to your suggestions for each Mathematical Practice.

> You should consider collecting this worksheet tool from every team in your area of school leadership and somehow posting the expectations for student learning in your classrooms. If your state or district uses a different set of mathematics process standards then do this same exercise for those specific and essential learning standards.

Observing the Effective Use of Higher-Level-Cognitive-Demand Tasks During the Unit of Instruction

Teachers' effective use of higher-level-cognitive-demand mathematical tasks during the unit of instruction (to develop student reasoning and problem-solving ability) will also benefit from collaborative team discussions. Teams can use the questions and prompts in figure 2.3 (page 68) to generate an effective discussion.

During the unit, team members should bring a higher-level-cognitive-demand task—one they selected from a resource or developed as part of HLTA 2 (page 20) from chapter 1 of this handbook. These tasks may take the full mathematics class time or period or multiple days, or they can take just ten minutes of class time. Just make sure that the tasks teachers use provide a higher-level-cognitive-demand expectation for student reasoning.

In general, when observing whether teachers use higher-level-cognitive-demand classroom tasks, you can use figure 2.3 (page 68) to monitor teacher action and intent. Think of the questions as potential look-fors when observing the mathematics class and as questions you would ask the teacher after your class visit.

Directions: Record your insights about proficiency with these Standards for Mathematical Practice and processes.

Mathematical Practice 1: When students are proficient with making sense of problems and persevering in solving them, they . . .

Mathematical Practice 2: When students are proficient with reasoning abstractly and quantitatively, they . . .

Mathematical Practice 3: When students are proficient with constructing viable arguments and critiquing the reasoning of others, they . . .

Mathematical Practice 4: When students are proficient with modeling with mathematics, they . . .

Mathematical Practice 5: When students are proficient with using appropriate tools strategically, they . . .

Mathematical Practice 6: When students are proficient with attending to precision, they . . .

Mathematical Practice 7: When students are proficient with looking for and making use of structure, they . . .

Mathematical Practice 8: When students are proficient with looking for and expressing regularity in repeated reasoning, they . . .

Figure 2.2: Identifying proficiency with the Standards for Mathematical Practice.

Visit **go.solution-tree.com/mathematicsatwork** to download a reproducible version of this figure.

Directions: Use these questions to allow for deep discussion of each higher-level-cognitive-demand mathematical task.

1. Why did you select or create this task?

2. How would you modify the task to increase its potential to engage students?

3. What is the targeted learning standard for this task?

4. What is the real-life topic of this task? What other real-life topic might be useful as a context for the same mathematics in the task?

5. What can you learn about students as they engage in and respond to the task?

6. How might you provide scaffolding to a student who struggles without lowering the cognitive demand of the task too much?

7. What do you expect students' wonderings and questions to be regarding the task?

8. What do you expect students' misunderstandings to be regarding the task?

9. What might make the mathematics of the task less or more challenging to meet the needs of students?

10. How do you propose assessment of the task should be effectively managed and facilitated?

Figure 2.3: Mathematics instruction and higher-level-cognitive-demand mathematical task development tool.

Visit **go.solution-tree.com/mathematicsatwork** to download a reproducible version of this figure.

Effective higher-level-cognitive-demand tasks provide an opportunity for students to showcase their learning in multiple ways. The way teachers manage students through the task involves important lesson-design decisions as they make sure students have the opportunity to develop their proficiency in the Mathematical Practices, such as with Mathematical Practice 1—"Make sense of problems and persevere in solving them."

To empower students to benefit from higher-level-cognitive-demand tasks, and to subsequently develop proficiency in making sense of problems and persevering in solving them, students need differentiated in-class support from the teacher. The teacher should guide and facilitate in-class problem solving as part of a peer-to-peer, engaging student experience. Figure 2.4 provides actions you can expect from teachers who provide differentiated support (scaffolding or advancing prompts) around meaningful mathematics tasks.

When observing teachers in the mathematics classroom, look for the following.
Are your teachers: <ul><li>Engaged in higher-level-cognitive-demand tasks with students during class so they can observe effective and expected problem-solving behavior? Students need constant feedback from the teacher and from each other during the learning experience the task generates.</li><li>Encouraging students to persist on a task, scaffolding as needed to support students' learning? Remember that the expectation is not needless student struggle but student productive struggle. Unproductive struggle is characterized by a situation in which students do not make progress toward sense making (Hiebert & Grouws, 2007; Warshauer, 2011).</li><li>Pulling from a pool of carefully selected hints or scaffolding prompts for higher-level-cognitive-demand tasks so that students can receive support to respond to a task without being given so much information that they do not need to put forth much effort (assessing questions and advancing questions)?</li><li>Helping students notice the progressions of structures in the mathematics content? This will help students to better recognize the types of differences and similarities between mathematical situations.</li><li>Helping students to embrace their errors and understand the value of learning from mistakes?</li></ul>

Figure 2.4: Teacher actions that provide targeted and differentiated support.

Visit **go.solution-tree.com/mathematicsatwork** to download a reproducible version of this figure.

To help your teachers plan for these mathematical task expectations, you can use the prompts in figure 2.5 (page 70) to provide differentiated in-class support for higher-level-cognitive-demand tasks and productive struggle. Figure 2.5 is also a support tool you can use to monitor and provide feedback to your teachers as part of your classroom observations. Teacher discussions around the provision of differentiated support will help each team avoid the temptation to *teach by telling*. You will need to ensure teachers no longer solely *teach by telling*, particularly when students are challenged to learn the mathematics via the personal experience of a higher-level-cognitive-demand task. Additionally, in either a pre- or post-conference situation, ask your teachers to respond to the questions in figure 2.5. Also, your collaborative teacher teams can use this tool when discussing their lesson designs during the unit.

Directions: Within your collaborative teams, have members complete each statement individually. Then, collectively discuss and decide on one to two indicators for differentiated support for students to engage in higher-level-cognitive-demand mathematics tasks.

1. My lessons allow students to take risks with problem-solving tasks and higher-level-cognitive-demand tasks because . . .

2. Students know I have high expectations for their development and perseverance as good problem solvers because . . .

3. The types of questions that build students' proficiency, perseverance, and endurance with our higher-level-cognitive-demand problem-solving tasks are . . .

4. When students are stuck on a problem . . .

   a. I do this . . .

   b. They do this . . .

5. If students need a resource or a model for a problem-solving task, they go to . . .

6. My feedback to students during a task is mostly based on effort by using phrases such as . . .

7. My feedback to students during a task is mostly based on talent by using phrases such as . . .

Figure 2.5: Prompts for differentiated in-class support and student perseverance.

Visit **go.solution-tree.com/mathematicsatwork** to download a reproducible version of this figure.

As a reminder, Carol Dweck's (2008) research provides critical insight into the nature of teacher praise. Students will be increasingly more apt to tackle and persevere through more challenging mathematical tasks when the nature of the praise they receive from teachers rewards their effort in doing the task, and not their "fixed ability" or "smarts" for getting the correct answer.

Observing a Sustained Focus on the Essential Learning Standards During the Unit of Instruction

During instruction, your collaborative teams should make inferences from students' engagement about the level of proficiency students are developing in select Mathematical Practices. Forward planning to provide differentiated support during the unit of instruction is an effective means to ensure your teachers are equipped with management strategies to guide students in the learning process.

However, it is important to remember that student proficiency in each of the Mathematical Practices—through the use of higher-level-cognitive-demand tasks—is best understood as a tool that *serves* the essential learning (content) standards of the unit. Otherwise, the mathematical task or problem a teacher presents to the students will seem random at best and disconnected at worst.

Consider the higher-level-cognitive-demand mathematical task in figure 2.6 (page 72).

Figure 2.7 (page 73) presents teacher strategies for maintaining the lesson's higher-level-cognitive-demand tasks and for ensuring students understand the essential learning standard each task addresses. Notice how the essential learning standard (8.G.3) requires a student to *describe*. Thus, the task needs to ensure the standard's expectations are served. Teachers can use the strategies in figure 2.7 to determine how well their task (such as figure 2.6) meets the essential learning standard expectations.

8.G.3: I can describe the effect of dilations, translations, rotations, and reflections on two-dimensional figures using coordinates.

Directions: Use the following diagram to answer the questions.

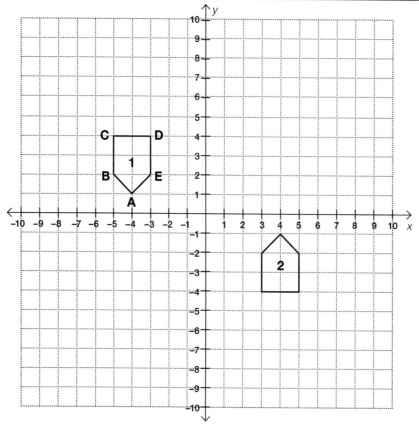

2a. Describe a series of transformations (reflection, rotation, translation, or dilation) that proves figure 1 is congruent to figure 2.

2b. Justify your answer to 2a.

2c. If you do a *translation* followed by a *reflection*, does it still map figure 1 to figure 2? How do you know?

Source: Adapted with permission from Aptakisic Junior High School, Buffalo Grove, Illinois.
Source for standard: NGA & CCSSO, 2010, p. 56.

Figure 2.6: Sample transformations using the coordinate grid.

Visit **go.solution-tree.com/mathematicsatwork** to download a reproducible version of this figure.

Strategies for teachers to maintain higher-level cognitive demand include the following.

- Motivate the necessity for task completion and solution pathways by connecting the task to the essential learning standard for the unit. (Label the essential learning standard as an "I can" statement at the top of the task.)

- Monitor students' initial responses to the tasks or problem. If students can easily find the solution, then the task most likely does not rise to the level of higher-level cognitive demand.

- Ensure the tasks encourage a process of reflection and communication.

- Present mathematical tasks that require students to apply mathematical knowledge and skills and are nonroutine. The goal is for students to translate their mathematical understandings from one context or representation to another as one indicator of mathematical proficiency.

- Rather than focus only on asking students to provide examples of mathematical concepts, request students to also provide counterexamples, and engage students in comparing and contrasting mathematical concepts.

- Engage students in solving problems orally, encouraging the important and proper use of mathematical vocabulary (precision of language) and logical sequencing of processes.

Figure 2.7: Strategies for maintaining higher-level-cognitive-demand tasks during instruction.

Visit **go.solution-tree.com/mathematicsatwork** to download a reproducible version of this figure.

Developing student proficiency with each of the Standards for Mathematical Practice and processes through the use of higher-level-cognitive-demand tasks is possible when your collaborative teams have a good understanding of each Mathematical Practice or process expected, have reached agreement on proficiency with each Mathematical Practice or process, and are committed to maintaining higher-level cognitive demand during instruction.

> You should ask your teachers three questions during a mathematics class observation or walkthrough.
>
> 1. "What is the essential *content* standard?"
>
> 2. "What is the essential *process* standard?"
>
> 3. "What is the *cognitive demand* of the mathematical tasks students will experience today?"

How students might engage with the mathematics and demonstrate their learning during instruction should become a routine part of your teams' work together.

Team Progress

It is helpful for collaborative teams to diagnose their reality and assess their actions for the effective use of higher-level-cognitive-demand tasks for student learning and demonstrations of understanding mathematics via the Mathematical Practices. Ask each team to use the status check tool for HLTA 6, using higher-level-cognitive-demand mathematical tasks effectively, in table 2.1 (page 74) to determine how well it is currently pursuing this high-leverage team action.

Table 2.1: During-the-Unit Status Check Tool for HLTA 6—Using Higher-Level-Cognitive-Demand Mathematical Tasks Effectively

Directions: Discuss your perception of your team's progress on the sixth high-leverage team action—using higher-level-cognitive-demand mathematical tasks effectively. Defend your reasoning.			
Stage I: Pre-Initiating	**Stage II: Initiating**	**Stage III: Developing**	**Stage IV: Sustaining**
We do not attend to or discuss Mathematical Practices or processes.	We have discussed Mathematical Practices and processes.	We consistently discuss the intent, purpose, and evidence of Mathematical Practices and processes during each unit of study.	We engage, as a team, in deep planning for Mathematical Practices and processes in our lessons.
We do not have a clear understanding of each Mathematical Practice.	We do not have a collaborative agreement on the focus of the Mathematical Practices for each unit of the course.	We teach some aspect of various Mathematical Practices as part of every daily lesson.	We use higher-level-cognitive-demand tasks as an intended activity to meet both the essential learning standards as well as Mathematical Practices and processes outlined for the lesson and the unit.
We do not use common higher-level-cognitive-demand tasks in order to develop students' Mathematical Practices.	We discuss and use some common higher-level-cognitive-demand tasks in class.	We discuss and implement collaboratively developed higher-level-cognitive-demand tasks.	We discuss and use intentional and targeted differentiated in-class supports as students engage in Mathematical Practices and processes by using our common and higher-level-cognitive-demand mathematical tasks.
We do not know the Mathematical Practices expected or demonstrated by the students assigned to other members of our team.	We have not reached team agreement on how to implement and sustain student proficiency in the Mathematical Practice expectations.	We do not collaboratively plan for Mathematical Practices, and they do not influence daily instructional plans for the unit.	We collaboratively plan for and implement Mathematical Practices and processes as part of our daily instructional plans for the unit.

Visit **go.solution-tree.com/mathematicsatwork** to download a reproducible version of this table.

HLTA 7: Using In-Class Formative Assessment Processes Effectively

[Formative assessment processes] can essentially double the speed of student learning producing large gains in students' achievement.

—W. James Popham

This high-leverage team action is where your leadership regarding mathematics instruction is critical. Once again, recall there are four critical questions every collaborative team in a PLC asks and answers on a unit-by-unit, ongoing basis.

1. What do we want all students to know and be able to do? (The essential learning standards)

2. How will we know if they know it? (The assessment instruments and tasks teams use)

3. How will we respond if they don't know it? (Formative assessment processes for intervention)

4. How will we respond if they do know it? (Formative assessment processes for extension and enrichment)

The seventh high-leverage team action—using in-class formative assessment processes effectively—ensures your collaborative teams reach clarity on how to effectively respond in class to the third and fourth critical questions of a PLC: How will we respond if they don't know it? How will we respond if they do know it? What will the teacher and teacher teams' formative responses be when students *are not* learning the standard for that day? What will be their in-class formative response when students *are* learning the standard for that day?

High-Leverage Team Action	1. What do we want all students to know and be able to do?	2. How will we know if they know it?	3. How will we respond if they don't know it?	4. How will we respond if they do know it?
During-the-Unit Action				
HLTA 7. Using in-class formative assessment processes effectively	◧	◧	▦	▦

▦ = Fully addressed with high-leverage team action

◧ = Partially addressed with high-leverage team action

The What

─────────────── **Real-Life Leadership Scenario** ───────────────

Structures often help move culture. This became clear in our work with a Wisconsin middle school. After two days of watching the school's mathematics teachers teach, we asked the teachers and the principal to tell us in thirty seconds or less their vision for mathematics instruction in their school. They struggled with the question. From

> *what we had observed, their vision for instruction appeared to be that the teacher does most of the work, and the students watch them do most of the work. We then asked the teachers on the collaborative team if they knew the difference between checking for understanding and using formative assessment processes. They knew the terms but were not exactly sure of their meanings.*
>
> *We asked them why their student desks were in rows. They did not know other than that was how it always was. We then mentioned that we had eliminated the use of rows in other school district classrooms more than twenty years before, and it was one of the primary reasons student performance in mathematics had increased so significantly. Sometimes a structural change is needed to create a cultural change, such as the peer-to-peer student discourse required in mathematics class.*
>
> *The structural change was not a sufficient factor for ensuring high-quality student discourse, but it was a necessary factor. We were all going to need to work together to discover how expert mathematics teachers manage small-group discourse really well.*

You should expect this seventh high-leverage team action to be a priority in your teams' discussions and professional development work as it supports student proficiency in the use of effective formative assessment processes in class. Typically, one might not think of assessment as a support for learning, but effective formative assessment is at the core of the PLC teaching-assessing-learning cycle (figure 2.1, page 62) and is critical to the student learning process (Kanold, Kanold, & Larson, 2012).

According to Dylan Wiliam (2011):

> When formative assessment practices are integrated into the minute-to-minute and day-by-day classroom activities of teachers, substantial increases in student achievement—of the order of a 70 to 80% increase in the speed of learning are possible. . . . Moreover, these changes are not expensive to produce. . . . The currently available evidence suggests that there is nothing else remotely affordable that is likely to have such a large effect. (p. 161)

Do not let your collaborative teams ignore this wise advice. Used effectively, this high-leverage team action will have a significant and positive impact on student learning in your school. This is what Hattie (2012) meant when he stated, "Teacher, know thy impact."

During the unit of instruction, part of your work is to ensure each grade-level or course-based collaborative team builds confidence in its use of in-class formative assessment *processes*. This process is much more than just observing evidence of student learning (checking for understanding), which is at best a diagnostic event.

For the teaching process to also be formative, teachers and student peers must provide meaningful and formative feedback to each other during engagement with the mathematics tasks and problems each day.

According to Douglas Reeves (2011) and Hattie (2009, 2012), there are four boundary markers that provide a basis for effective formative feedback to students. We have organized these markers into the acronym FAST.

1. **Feedback must be _Fair_:** Does teacher feedback rest solely on the quality of student work and not on other characteristics of the student, including some form of comparison to others in the classroom?

2. **Feedback must be _Accurate_:** Is the feedback during the in-class activity actually correct? Do students receive prompts, solution pathway suggestions, and discourse that are effective for understanding the mathematical task as the teacher tours the room observing students during peer-to-peer discourse, checking for understanding, and providing feedback?

3. **Feedback must be _Specific_:** Does the verbal feedback students receive contain enough specificity to help them persevere and stay engaged in the mathematical task or activity process? Does the feedback help students get unstuck or advance their thinking as needed? (For example, "Work harder on the problem" is not helpful feedback for a student.)

4. **Feedback must be _Timely_:** Does the teacher provide immediate and corrective feedback to keep students on track for the solution pathway as he or she tours the room, observes students during peer-to-peer discourse, and listens in on the peer-to-peer conversations?

However, FAST meaningful feedback from teacher to students or from student peers to each other will still not be sufficient for student learning. If, during the best teacher-designed moments of classroom formative assessment processes, teachers fail to support students *taking action* on evidence of continued areas of difficulty (step three of the PLC teaching-assessing-learning cycle, page 62), then the cycle of learning, assessing, and continued learning stops for the student (Wiliam, 2011).

The How

According to W. James Popham (2011), when teachers use formative assessment well:

> It can essentially double the speed of student learning producing large gains in students' achievement; and at the same time, it is sufficiently robust so different teachers can use it in diverse ways and still get great results with their students. (p. 36)

So what does this have to do with the work of your collaborative teams? Your work should begin with making sure each team is clear about the difference between the often used (and misused) in-class teaching techniques of checking for understanding and using formative assessment processes.

The Difference Between Checking for Understanding and Using Formative Assessment Processes

Your teachers need to be clear with students about the difference between checking for understanding in class (which may be helpful to the teacher but has minimal impact on student learning) and using formative assessment processes in class (which includes the classroom elements of formative feedback and student action on feedback).

To facilitate team discussions of these differences, your teams should engage in the checking for understanding versus using the formative assessment process tool in figure 2.8 (page 78).

Directions: Answer the following questions.

1. How do you currently check for student understanding on a daily basis? Write your responses in the Checking for Understanding column. For each response, indicate whether you use whole-group discourse (teaching at the front of the room) with a *W*, small-group discourse (students working with their peers) with an *S,* or independent practice (students working individually) with an *IP.*

2. What do you and your colleagues believe is the difference between solely checking for understanding and using formative assessment processes in the classroom during instruction?

3. How can you implement formative assessment processes in the classroom each day?

Remember, for the process to be formative, students must actually take action on the feedback they receive. In the Using Formative Assessment Processes column, explain ways that you could improve each check for understanding you listed to become a moment of formative assessment for your students.

Checking for Understanding	Using Formative Assessment Processes

Figure 2.8: Checking for understanding versus using the formative assessment process tool.

Visit **go.solution-tree.com/mathematicsatwork** to download a reproducible version of this figure.

At the heart of a robust vision for mathematics instruction is what students are *doing* during class. How are they engaged? With whom are the most significant conversations taking place in the classroom: student-to-student or teacher-to-student? Moreover, in your school, do students see each other as reliable and valuable resources for learning?

Hattie and Gregory Yates (2014, p. 49) refer to the idea that our lessons should be designed to encourage the student voice:

> Attempts have been made to alter classroom procedures to encourage quality student talk and active teacher listening. . . . [R]educing teacher talk and increasing student talk was not enough. Student talk is a means to an end. Deliberate strategies are needed to structure student talk that leads to greater comprehension and learning.

Teachers promote conceptual understanding when they explicitly make, or ask students to make, connections among ideas, facts, and procedures (Hiebert & Grouws, 2007). The following five teacher actions help students make these connections (Kanold, Briars, & Fennell, 2012).

1. Challenging students to think, reason, and make sense of what they are doing to solve mathematics problems
2. Posing whole-group and small-group questions that stimulate students' thinking and reasoning and asking them to justify their conclusions, solution strategies, and processes
3. Expecting students to evaluate and explain the work of other students and engage in peer-to-peer discourse by comparing and contrasting different solution pathways for the same problem, while receiving formative feedback on the accuracy of those pathways
4. Expecting students to engage in productive struggle and action necessary to find successful solutions to the tasks
5. Asking students to represent the same ideas in multiple ways—using multiple representations, such as numerical tables, graphs, or models to demonstrate understanding of a concept

Good mathematics instruction places a lot of the teachers' instructional energy into using and managing small-group discourse and providing opportunities for meaningful formative feedback to students during class (NCTM, 2014). However, for feedback to be effective, students must take action on the formative feedback they receive either from the teacher or from their peers.

As a point of practice, when you observe a mathematics class, your focus should be more on what the students are doing and less on the teacher. Observations of students learning mathematics should reveal about 60–65 percent of the class time as well-managed peer-to-peer student discourse.

Thus, going beyond checking for understanding from the front of the classroom and moving into the more beneficial formative feedback process as part of instruction require your teachers to move away from long periods of lecture or demonstration at the front and into the use of rich mathematical tasks that support more robust student (peer-to-peer) discourse. It also requires well-managed activities as teachers loop through class and orchestrate all types of scaffolding and advancing prompts, supporting students in productive and engaged exploration and discussion with their peers.

The Heart of Formative Assessment: Student Reasoning, Discourse, and Engagement Using Differentiation of Tasks

Consider Mathematical Practice 2—"Reason abstractly and quantitatively." One of the primary outcomes of the mathematics standards is the expectation students will experience *reasoning* during the mathematics lesson every day in the classroom. Do your collaborative teams' discussions help each of your teachers prepare and plan for an expectation of student reasoning every day? How do you know?

More important, what does *to reason* mean? What should you expect students to be doing if you are observing their reasoning? Do students reason alone? With the teacher? With each other? Out loud? In debate? On paper or with technology? The answer is yes to all of these methods of demonstrating reasoning. The higher the cognitive demand of the mathematical tasks teachers present for *students* to do in class (not for them to watch the *teacher* do), the greater the levels of complexity of reasoning students will display. Teacher commitment (or lack thereof) to ensuring that reasoning takes place every day will determine the kinds of mathematical experiences the teacher uses to assess student understanding. If teacher commitment varies widely in an individual teacher's pedagogical approach, then inequities will persist in student learning as well.

Consider the grade 8 higher-level-cognitive-demand reasoning task in figure 2.9.

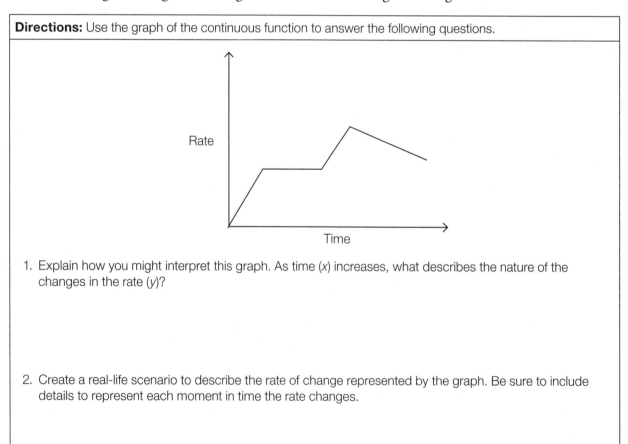

Directions: Use the graph of the continuous function to answer the following questions.

1. Explain how you might interpret this graph. As time (*x*) increases, what describes the nature of the changes in the rate (*y*)?

2. Create a real-life scenario to describe the rate of change represented by the graph. Be sure to include details to represent each moment in time the rate changes.

Figure 2.9: Grade 8 higher-level-cognitive-demand reasoning task.

Visit **go.solution-tree.com/mathematicsatwork** to download a reproducible version of this figure.

With higher-level-cognitive-demand questions such as the grade 8 reasoning task, your teachers should anticipate the type of reasoning and solution pathways they might expect students to pursue when engaging in the task during class.

Differentiation will occur as the teacher walks around the room and provides feedback to the student teams. Some students might need some scaffolding prompts to stay engaged on the task; for example, some students may not clearly understand what it means to describe the "nature of the changes."

All students would be expected to work the task together in their small groups. More advanced student teams could also reason through an extension to the task such as, "If someone were to walk at the rate of change shown in this graph, describe the nature of both the direction and the speed of the path."

In and of itself, a higher-level-cognitive-demand reasoning task like figure 2.9 does not guarantee student reasoning and understanding will take place, however:

> Accomplished teachers deliberately structure opportunities for students to use and develop appropriate mathematical discourse as they reason and solve problems. These teachers give students opportunities to talk with one another, work together in solving problems, and use both written and oral discourse to describe and discuss their mathematical thinking and understanding. As students talk and write about mathematics—as they explain their thinking—they deepen their mathematical understanding in powerful ways that can enhance their ability to use the strategies and thought processes gained through the study of mathematics to deal with life issues. (National Board for Professional Teaching Standards, 2010, p. 57)

The manner in which your teachers facilitate discourse while using the task in class is critical to creating and supporting a classroom culture and learning environment that value reasoning and sense making and are research-informed instructional practices (NCTM, 2014). Your teachers decide what thinking to share and whose voices are heard, which has a profound impact on knowledge creation. "Teachers, through the ways in which they orchestrate discourse, convey messages about whose knowledge and ways of thinking and knowing are valued, who is considered able to contribute, and who has status in the group" (NCTM, 1991, p. 20).

Small-group discourse with a student team can also help your teachers meet the expectations for Mathematical Practices, such as Mathematical Practice 3, "Construct viable arguments and critique the reasoning of others." Discourse among small groups of students opens a window into student reasoning about mathematics:

> Students who learn to articulate and justify their own mathematical ideas, reason through their own and others' mathematical explanations, and provide a rationale for their answers develop a deep understanding that is critical to their future success in mathematics and related fields. (Carpenter, Franke, & Levi, 2003, p. 6)

When you observe a mathematics class, you should *see and hear* the connections students make, the questions they ask each other as well as the teacher, and the obstacles or misconceptions that can hinder their conceptual understanding. Teachers are in a better position to make decisions and implement strategies to support learning and push the level of reasoning and problem solving expected in Mathematical Practice 1, "Make sense of problems and persevere in solving them," through the use of small-group student dialogue. This dialogue creates a community of student learners who collectively build mathematical knowledge and proficiencies *together*.

You can use the student team observation tool in figure 2.10 to collect information and provide feedback to your teachers on what you observe when student teams are engaged in the mathematical tasks and activities teams planned for class. You could also use this tool for data collection during a classroom observation walkthrough or visit.

When observing this formative feedback process in class, you should notice the teacher providing guidance and scaffolding questions to support student learning and perseverance on the task. Ask your collaborative teams to discuss how this tool or a modification of it might be useful in providing information about students' development of proficiency to reason abstractly and quantitatively during mathematics lessons.

Do the students in your school learn to gradually take more responsibility for their mathematics learning by reflecting on their in-class work (with their peers) long enough to view solution mistakes as learning opportunities? This is essential for student learning to become part of a formative process in your school. As Dylan Wiliam (2007) indicates, in order to "improve the quality of learning within the system, to be formative, feedback needs to contain an implicit or explicit recipe for future action" (p. 1062).

When students make errors, they need to receive feedback on those errors, and then take action to correct their mistakes in reasoning. Students then view assessment of their work as something they *do* in order to focus their energy and effort for future learning. As indicated so far in this section, a great place for this type of student reflection is during small-group discourse as mathematics students work together on various problems or tasks while the teacher walks around the room, checks for understanding, and then provides meaningful feedback. Thus, teachers turn checks for understanding into formative assessment processes.

An important component, then, for supporting student proficiency in the Mathematical Practices through the use of in-class higher-level-cognitive-demand tasks is to ask your teachers to be explicit with their students and their plans to implement effective formative assessment strategies during class.

Management of Students in Peer-to-Peer Productive Discourse

Students and teachers share the responsibility to successfully implement the in-class formative assessment processes and practices. When students can demonstrate understanding through reasoning mathematically, they connect to the essential learning standard for the unit and can reflect on their individual progress toward that learning standard. Teachers support students' progress by using immediate and effective feedback during the daily classroom conversations, whether during whole-group or small-group moments. The best structural vehicle to observe student understanding, provide feedback, and help students to take action on that feedback is small-group student discourse. This means students should no longer be sitting in rows during instruction.

Managing students in peer-to-peer productive discourse is critical for effective implementation of lower- and higher-level-cognitive-demand tasks. When your teachers set student teams to work, do they know the expectation for teamwork, or do they rely on one or two students to get them started? Your students will need structure to support meaningful mathematics discourse, engagement, and action.

Student Teams (List Team Member Names)	Level 1 Team is working the task but is constantly stuck. Students do not generally take corrective action on the teacher's feedback and scaffolding prompts.	Level 2 Team is working the task by connecting to prior knowledge, engaging in conversations, and taking corrective action to the teacher's scaffolding prompts.	Level 3 Team is working the task by engaging in accurate sense making and reasoning and using multiple connections with minimal teacher feedback.	Level 4 Team has worked the task and completed it correctly and is engaging in an extension to the problem (advancing prompts) provided by the teacher.
Student team 1				
Student team 2				
Student team 3				
Student team 4				
Overall observations of student reasoning and engagement in the higher-level-cognitive-demand task:				

Figure 2.10: Walkthrough formative assessment observation tool.

Visit **go.solution-tree.com/mathematicsatwork** to download a reproducible version of this figure.

As students are engaged in the mathematics, structures of the engagement help to set the expectations for teamwork. Students should know the expectations for team interactions and their rights and responsibilities to their team. Your collaborative teacher teams should create posters to share with you and their students about expectations for behavior when working with other student peers. See figure 2.11 for an example.

Working in Teams

Rights

Each student has the right to:

- Have a voice in discussions
- Ask for assistance
- Embrace mistakes
- Express his or her opinion
- Learn from his or her team
- Disagree with respect
- Learn rich mathematics

Responsibility

Each student should:

- Actively listen to all team members
- Help others when asked
- Provide positive feedback
- Be encouraging and open to other perspectives
- Respect others' rights
- Seek consensus
- Ensure success for all team members

Source: Adapted from South Mountain High School, Phoenix, Arizona, Mathematics Department.

Figure 2.11: Team discourse rules and responsibilities.

Once students understand the expectations for team behaviors, teachers can incorporate structure to rich mathematical tasks so all members have a role in the task. For example, there are several cooperative learning structures teachers can employ to create required participation. The key is to engage students in mathematical thinking (Johnson & Johnson, 1999; Johnson, Johnson, & Holubec, 2008; Kagan, 1994; Kagan & Kagan, 2009).

In general, small-group discourse classroom moments should have the following five key elements (Johnson & Johnson, 1991; Kagan, 1994; Kagan & Kagan, 2009).

1. **Positive interdependence:** Teachers should structure the task so group members must work together to complete the task. Each team member will have a unique contribution to the team, and team members rely on each other to complete the task.

2. **Individual accountability:** All students in the group must be accountable for completing the task.

3. **Face-to-face feedback and interactions:** Students will be engaged in the task by providing feedback to each other, negotiating solution strategies, taking risks, testing each other's conjectures, and ensuring that everyone on the team understands the team's thinking.

4. **Interpersonal and small-group relational skills:** Team members use appropriate skills to support each other, develop trust, and build a community of learners.

5. **Team processing:** Students self-assess how they are working together, and teachers provide feedback to teams on their ability to work cooperatively.

As your collaborative teams plan learning experiences, they can use figure 2.12 to list how the activity promotes these five key elements of effective small-group student discourse.

Essential learning standard: _____

Small-group discourse learning activity: _____

Directions: Indicate how this learning activity promotes each of the following discourse criteria for each student team.

Positive Interdependence	Individual Accountability	Face-to-Face Feedback and Interactions	Interpersonal and Small-Group Relational Skills	Team Processing

Figure 2.12: Small-group discourse peer-to-peer engagement tool.

Visit **go.solution-tree.com/mathematicsatwork** to download a reproducible version of this figure.

For more explicit information on how teachers can manage peer-to-peer small-group discourse in their mathematics classes, read *Five Practices for Orchestrating Productive Mathematics Discussions* (Smith & Stein, 2011). However, it is advisable to help your teachers be very clear with structured directions to the students regarding whole-group versus small-group discourse aspects of the class. Students' attention deteriorates over the course of the mathematics lesson (Hattie & Yates, 2014). This can be avoided through the use of small-step instruction moving frequently in and out of small-group versus whole-group discourse.

Formative Assessment and Tier 1 RTI Differentiated Responses to Learning

Douglas Fisher, Nancy Frey, and Carol Rothenberg (2010) suggest that "interventions are an element of good teaching" (p. 2), and these interventions begin in the classroom. In-class Tier 1 RTI is the first line of defense for struggling students and includes a differentiated response to learning each mathematical task as your teachers provide students and student teams with scaffolding prompts to help them think of other ways to solve a problem or higher-level-cognitive-demand task.

Within the RTI framework, teacher teams purposefully and explicitly plan differentiation. The use of formative assessment data and knowledge of students' language, diverse culture, and prior knowledge offer students in the same class different teaching and learning opportunities to address their learning needs—especially as teachers give them feedback on their progress and they take action with their peers. By using formative assessment strategies, teachers are not making the content easier; they are making the content more accessible by the time the class period ends.

Teachers' willingness and adaptability to try different teaching strategies and solution pathways to help all students understand the learning of the standard is in and of itself a Tier 1 response. No matter how great the lesson, if it doesn't result in student demonstrations of understanding, then seeking other ways to teach the standard is a great Tier 1 teacher-led response.

You can change the culture of mathematics lessons in your school by expecting (requiring) small-step instruction—that is, students moving between whole-group (teacher led) discourse and small-group (peer to peer) discourse. After ten minutes of teacher lecture, student attention deteriorates into cascading inattention and mind wandering. Without constant feedback and action, there is no understanding (Hattie & Yates, 2014).

Team Progress

It is helpful for your collaborative teams to diagnose their current reality and action during the unit. Ask each team to assess its progress on the seventh high-leverage team action, using in-class formative assessment processes effectively, using the status check tool in table 2.2. As your teams seek stage IV—sustaining—they will increase rigor, coherence, and fidelity toward student improvement and demonstration of proficiency in various Mathematical Practices and processes your teams focus on during the units of instruction. Teachers will also support deeper levels of student understanding for the learning standards, increasing students' chances for success on end-of-unit assessments.

So much of student learning depends on the decisions your teachers make every day. Will they use a robust set of mathematical tasks? Will *every teacher* become intentional about using effective in-class formative assessment processes that exhibit meaningful feedback with subsequent student action? And how will you know? Will there be a great push of more small-step instruction and student talk in class every day? It is a lot to ask.

To help you become more organized in how you might think through daily mathematics instruction, this chapter concludes with the eighth high-leverage team action—using a lesson-design process for lesson planning and collective team inquiry. All of the elements of research-affirmed instruction come together in this final section of chapter 2.

Table 2.2: During-the-Unit Status Check Tool for HLTA 7—Using In-Class Formative Assessment Processes Effectively

Directions: Discuss your perception of your team's progress on the seventh high-leverage team action—using in-class formative assessment processes effectively. Defend your reasoning.

Stage I: Pre-Initiating	Stage II: Initiating	Stage III: Developing	Stage IV: Sustaining
We do not attend to or discuss formative assessment processes used in our instruction.	We have discussed formative assessment processes and do not need to do so again.	We emphasize to students the value of using formative assessment feedback during class and taking action during class.	We always use in-class formative assessment processes to inform students' reasoning and learning.
We do not have a clear understanding of the difference between checking for understanding and formative assessment processes.	We discuss and use checking for understanding methods in class, but do not provide feedback or expect action.	We plan for student reasoning and sense making, and it is built into a formative feedback process in class.	We engage, as a team, in deep planning for formative assessment processes through the use of small-group discourse with teacher feedback and student action.
We do not know the in-class formative assessment methods used or expected by other members of our team.	We do not plan for how to engage students in formative assessment processes and practices as part of our team focus for each unit.	We plan for effective small-group student team activities as a way to promote the formative assessment process in class.	We discuss and implement well-developed formative assessment procedures to use with many of our higher-level-cognitive-demand tasks.
We do not use in-class formative feedback with student action in order to develop students' mathematical practice and learning.	We have not yet reached team agreement on how to use differentiated and targeted in-class Tier 1 RTI supports as part of our instruction.	We do not know the Tier 1 RTI response in class by other members of our team.	We discuss and use intentional differentiated and targeted in-class Tier 1 RTI supports as students engage in common higher-level-cognitive-demand mathematical tasks.

Visit **go.solution-tree.com/mathematicsatwork** to download a reproducible version of this table.

HLTA 8: Using a Lesson-Design Process for Lesson Planning and Collective Team Inquiry

Visible learning means an enhanced role for teachers, as they become evaluators of their own teaching.

—John Hattie

This eighth high-leverage team action highlights one of the most important activities you should support and participate in as a school leader. It is up to you to lead the visible learning of your teachers with one another. Your role is to expect and support transparent teacher instruction and practice. Your teacher teams must work to develop, discuss, use, and observe daily mathematics lessons that encourage students to think deeply about the learning standards and to demonstrate proficiency in the Mathematical Practices—a culminating practice critical to the adult learning in your school.

Recall there are four critical questions every collaborative team in a PLC asks and answers on a unit-by-unit, ongoing basis.

1. What do we want all students to know and be able to do? (The essential learning standards)

2. How will we know if they know it? (The assessment instruments and tasks teams use)

3. How will we respond if they don't know it? (Formative assessment processes for intervention)

4. How will we respond if they do know it? (Formative assessment processes for extension and enrichment)

High-leverage team action 8—using a lesson-design process for lesson planning and collective team inquiry—ensures your teams reach clarity on all four of the PLC critical questions within their mathematics lessons. The lesson-design process ensures all team members possess a strong understanding of the intent of the standard for the lesson, the purpose of the tasks to be used, and the expectation of student mastery of the standard.

High-Leverage Team Action	1. What do we want all students to know and be able to do?	2. How will we know if they know it?	3. How will we respond if they don't know it?	4. How will we respond if they do know it?
During-the-Unit Action				
HLTA 8. Using a lesson-design process for lesson planning and collective team inquiry	▢	▢	▢	▢

▢ = Fully addressed with high-leverage team action

The What

Real-Life Leadership Scenario

"She won't let us do it," they said. "Why not?" I [Tim] asked. "The district will never pay for the subs, and our principal won't push for it." I assured the sixth-grade math teachers that their principal was their advocate, and if we pushed for it, with good reason, she would remove the barrier. The teacher team members were not so sure.

We had created a mathematics lesson together—not the full-blown lesson-study type of lesson, just a rock-solid lesson on unit rates—that would get students to both reason and demonstrate precision of language as they worked on the tasks we created. We even had a brave teacher on the team volunteer to teach the lesson the next day during third hour. There was only one problem: almost all of the teachers on the team also needed to teach their own class during third hour. We needed subs!

So we went as a team to see the principal. We invited her to come observe the third-hour lesson with us and sit in on the debriefing meeting as well. We also informed her we would need everyone released from fourth hour for the debriefing. Could she make it happen? We promised this would be one of the greatest actions her sixth-grade math team could take this year, from a professional development and learning point of view. We also promised that, in the future, we would warn her at least four weeks in advance when we'd need subs. (We actually planned to do the lesson study process every nine weeks for the full year.) She was a bit hesitant.

She knew this could be a significant cultural event for this sixth-grade team and wanted to support them. These teachers had never observed each other's classes. It was time to become more transparent. The very idea they would openly visit each other with a designed academic learning purpose was foreign to them. This was their chance for a new beginning. And they were ready. Now they just needed permission and, in this case, scheduling support from the principal.

Effective mathematics instruction rests, in part, on the careful planning of your teacher teams (Morris, Hiebert, & Spitzer, 2009). Your grade-level collaborative teams are uniquely structured to provide time and support needed to interpret the essential learning standards, embed Mathematical Practices and processes into daily lessons, and reflect together on the effectiveness of your lesson implementation.

In this eighth high-leverage team action, your teachers should use lesson-planning protocols (a tool of some sort) during the unit instruction. One such tool (see figure 2.13, pages 91–92), adapted from the 2012 *Common Core Mathematics in a PLC at Work* series, provides a way to support the focus and design of teachers' mathematics tasks. It can help you and your teams plan for formative assessment feedback questions and evidence *during* the mathematics lesson. Consider the following three questions.

1. How do we expect students will express their ideas, questions, insights, and difficulties?

2. Where and when will and should the most significant conversations take place (student to teacher, student to student, or teacher to student)?

3. How approachable and encouraging should we be as students explore? Do students use and value each other as reliable and valuable learning resources?

What separates the Mathematical Practices lesson-planning tool from most other lesson-planning models is its focus on developing students' proficiency with Mathematical Practices *and* the daily essential content learning standards; its focus on what the teacher and students will be doing; and its focus on the beginning, middle, and end of the math class time.

This lesson-planning tool also specifically asks you and your team to consider both student and teacher actions during lesson planning, which helps your teachers understand learning through students' eyes and helps students become their own teachers. Thus, learning becomes more visible to you, your colleagues, and your students. One particularly powerful collaborative tool is lesson study. Lesson study is very effective as a collaborative protocol with a high impact on teacher professional learning (Gersten, Taylor, Keys, Rolfhus, & Newman-Gonchar, 2014; Hiebert & Stigler, 2000). Your teams can also use the Mathematical Practices lesson-planning tool as teachers participate in collective inquiry during the unit. In addition, the tool emphasizes the importance of selecting and using higher-level-cognitive-demand tasks along with the more common lower-level-cognitive-demand tasks to support students' mathematics learning.

The How

The work of your collaborative teams during the unit of instruction involves engaging one another in discussions about individual lessons. These lessons are focused on specific essential questions for the learning target that day.

The Mathematical Practices Lesson-Planning Tool

At this point, there are some aspects of the Mathematical Practices lesson-planning tool that your collaborative teams should contemplate to be successful. Consider the following frequently asked questions and responses about the Mathematical Practices lesson-planning tool.

Question: Is the Mathematical Practices lesson-planning tool appropriate for all mathematics content?

Answer: Yes. Each individual component of the tool is conducive to any mathematics content. In addition, during instruction for the unit, the structure this tool provides helps students come to know what to expect in a mathematics lesson and, hence, provides them some opportunity to note consistency and structure in doing the mathematics of the lesson.

Question: Will the Mathematical Practices lesson-planning tool work in a class of diverse learners?

Answer: Yes. When the Mathematical Practices lesson-planning tool asks, "What will the students be doing?", there is no focus on any particular characteristic of the student. In other words, the question addresses any student and every student—this includes consideration at the individual student level. So, the question is relative to your students and their characteristics. Other things, such as the structure of the classroom, also determine what students will be doing. For example, when students are working during the small-group instruction activities, what they are doing looks very different from students who are working individually or participating in discourse during whole-group instruction.

Unit:	Date:	Lesson:

Essential learning standard: List the essential learning standard for the unit addressed by today's lesson.

Learning objective: As a result of class today, students will be able to . . .

Essential standard for Mathematical Practice: As a result of class today, students will be able to demonstrate greater proficiency in which standard for Mathematical Practice?

Formative assessment process: How will students be expected to demonstrate mastery of the learning objective during in-class checks for understanding, teacher feedback, and student action on that feedback?

Probing Questions for Differentiation on Mathematical Tasks

Assessing Questions	**Advancing Questions**
(Create questions to scaffold instruction for students who are "stuck" during the lesson or the lesson tasks.)	(Create questions to further learning for students who are ready to advance beyond the learning standard.)

Tasks	**What Will the Teacher Be Doing?**	**What Will the Students Be Doing?**
(Tasks can vary from lesson to lesson.)	(How will the teacher present and then monitor student response to the task?)	(How will students be actively engaged in each part of the lesson?)
Beginning-of-Class Routines How does the warm-up activity connect to students' prior knowledge, or how is it based on analysis of homework?		

Figure 2.13: Mathematical Practices lesson-planning tool.

continued →

Task 1 How will the students be engaged in understanding the learning objective?		
Task 2 How will the task develop student sense making and reasoning?		
Task 3 How will the task require student conjectures and communication?		
Closure How will student questions and reflections be elicited in the summary of the lesson? How will students' understanding of the learning objective be determined?		

Source: Kanold, Kanold, & Larson, 2012, pp. 53–54.
Source for standards: NGA & CCSSO, 2010, pp. 6, 26.

Figure 2.13: Mathematical Practices lesson-planning tool (continued).

Visit **go.solution-tree.com/mathematicsatwork** to download a reproducible version of this figure.

Question: We know the use of effective formative assessment strategies requires the collaborative team to make a commitment to plan an in-class common formative assessment process. Where do we see evidence of this process in the Mathematical Practices lesson-planning tool?

Answer: Every point of the lesson provides the opportunity for teachers' assessment of student understanding and collection of information. For that data collection to become part of a formative assessment process for the student, every student must adjust and respond to his or her feedback and interaction with the teacher or their peers during the lesson.

As such, every point of your lesson should provide a student opportunity to experience an effective formative assessment process. For example, the tool's "Assessing Questions" section provides an opportunity for teachers to help students scaffold their learning on a task or lesson to persevere and make progress during the lesson and the unit.

Question: Is the Mathematical Practices lesson-planning tool compatible with any mathematics curriculum or textbook series?

Answer: Yes. The Mathematical Practices lesson-planning tool supports all mathematics instructional designs, so it is compatible with any mathematics curriculum or textbook series.

You can use the questions from figure 2.14 (page 94) to help support teacher study of the lesson-planning tool from figure 2.13.

Part of your role is to support and expect your collaborative teams to use either the Mathematical Practices lesson-planning tool or some other tool to plan lessons together at least once per nine weeks—if not more often. As they become confident in the high-leverage team actions both before and during the unit, they will find it easier to design, reflect on, plan, and implement more lessons *together*.

Lesson Study: Using the Mathematical Practices Lesson-Planning Tool for Collective Inquiry

In lesson study, your teacher teams develop an expectation related to student learning for the lesson. Teachers identify a particularly challenging mathematical standard for students and then together design a lesson to address a process for student learning of the standard. In the process of the lesson design, the team explores ideas about student learning as it relates to the chosen essential standard.

A more informal model for promoting the idea of lesson study and a collective inquiry into your teachers' instructional practice is for one or two collaborative team members to enter lesson element ideas into the lesson-planning tool and present it to the team for discussion and feedback. All team members subsequently contribute ideas and suggestions to the revision of the lesson. This is not the more formal Japanese lesson study model popularized in the 2000s (Fernandez & Yoshida, 2004), but rather an informal form of collective inquiry with other team members.

When possible, you should consider participating in and observing the lesson design with team members when one member volunteers to teach the lesson. You should also be part of the debrief discussion about observations of student action and learning during the lesson, make changes to the lesson design, and then support teachers in reteaching the lesson with a final debriefing of the second instructional episode. It may seem time consuming and be a bit intensive for every unit, but try to support some type of collective inquiry or lesson study at least once or twice per grading period per team.

Directions: Closely examine the Mathematical Practices lesson-planning tool (figure 2.13, pages 91–92). In your collaborative team, discuss how you can use the lesson-planning tool to help your current efforts to plan mathematics lessons for your grade level.

1. What are important factors to consider when planning lessons for an in-class formative assessment process?

2. How can you use the Mathematical Practices lesson-planning tool's assessing and advancing questions to facilitate small-group student-to-student discourse and differentiate learning around the higher-level-cognitive-demand mathematical tasks you use during the lesson?

3. How does the Mathematical Practices lesson-planning tool support the elements of your lesson—both the learning standards and the Mathematical Practices?

4. How will each mathematical task you choose for the lesson ensure students will be actively engaged in the mathematics?

5. Consider your start-of-lesson and end-of-lesson routines. How do these routines connect to the learning objectives of your lesson?

Figure 2.14: Learning about the Mathematical Practices lesson-planning tool.

Visit **go.solution-tree.com/mathematicsatwork** to download a reproducible version of this figure.

Our experience with teacher teams is this will be one of the most rewarding work activities they can do together. The benefit of a collective inquiry, lesson-study team activity is the teacher professional learning that results in the deep, collaborative discussions about content, student learning, and mathematics instruction. Note that you may need to broker a specific schedule or time during the normal work day to allow your teacher teams time to be able to do the lesson-study work required during the unit.

When thinking about a lesson-study lesson, remember a key benefit of the collaborative-planning process is to use it to develop teacher expectations for learning around Mathematical Practices or the process of how students learn during the lesson. For example, consider Mathematical Practice 1, "Make sense of problems and persevere in solving them." As the collaborative team plans the lesson, the discussion around how to promote perseverance in problem solving will foster team alignment of the content expectations *and* the performance expectations for the learning standard. By the end of the lesson study, teachers generally gain content knowledge and pedagogical knowledge. They will also raise the level of respect and trust among the other lesson-study team members.

You can use the tool in figure 2.15 to help each team with its collaborative planning and analysis for a lesson study. If you do so, you should be sure to collect the lesson design from the team and validate their work and effort.

Lesson-Design Components	Questions to Consider	Comments
Essential learning standard	What is the essential learning objective for the lesson? How does it connect to the essential learning standards for the unit?	
Assessment (formative, embedded, or summative)	What formative assessment strategies will we employ during the lesson? How will students self-assess their understanding and the understanding of their peers?	
Questioning	What are the assessing questions or prompts we can use to get students unstuck during the lesson? What are the advancing questions we can use to further student understanding? How will we engage students so that *all* students are required to think about the questions? How will we address a balance of cognitive-demand-level questions?	
Mathematical Practice and process	Which Mathematical Practice and process will students develop? How will they develop it? What evidence will we observe?	
Beginning-of-class routine	How do we address prior knowledge? How will all students be engaged in the opening activity?	
Activity or task 1: How will students be engaged in understanding the essential learning standards?	What are all the ways the task can be solved? Which of these methods do you think students will use? What misconceptions might students show? What errors might students make?	
Activity or task 2: How will the task develop student sense making and reasoning?	What is our expectation of student engagement for sense making and reasoning? How will we ensure the task is accessible to all students while still maintaining a high cognitive demand? What student-to-student interaction will we employ?	
Activity or task 3: How will the task require student conjectures and communication?	How does the plan address orchestrating the class discussion? Which solution paths do we want shared during the class discussion? In what ways will the order in which solutions are presented help develop students' understanding of the mathematical ideas that are the focus of the lesson?	
Student-led closure	How do students know that they met the learning target for the day? What evidence will we use to determine the level of student learning of the target?	

Source: Kanold, Kanold, & Larson, 2012, pp. 53–54.

Figure 2.15: Mathematical Practices lesson-study planning and analysis tool.

Visit **go.solution-tree.com/mathematicsatwork** to download a reproducible version of this figure.

You can then use the observation tool in figure 2.16 as a simple way to record data about what students are doing during the lesson that your teams prepare and observe. Remember that the intent of the lesson study, or collective inquiry as a team, is not so much to observe the teacher, as it is to observe the students and determine if they actually learned the standard via the tasks teachers chose for them to do that day. If possible, participate in the observation, note-taking, and debriefing activity.

Date: Course:	
Lesson Learning Standard: _____	
Lesson-Design Components	**Observations of Student Actions (Observed Data—What Are Students Doing?)**
Assessment (formative, embedded, and summative)	
Questioning	
Mathematical Practice and Process	
Beginning-of-Class Routine	
Activity or Task 1	
Activity or Task 2	
Activity or Task 3	
Student-Led Closure	

Figure 2.16: Lesson-study student data observation tool.

Visit **go.solution-tree.com/mathematicsatwork** to download a reproducible version of this figure.

Collaborative Planning and Collaborative Reflection

Collaborative planning is not the end of teachers' collaborative work on lesson design. An essential final step is collaboratively reflecting on the success of the lesson as implemented in collaborative team members' classrooms. Also important is what teachers can learn from each lesson to inform both upcoming lessons in future units as well as the design of a similar lesson for future implementation. Consider using the reflection and debriefing prompts in figure 2.17 for the debriefing session following the lesson observation.

Directions: With your collaborative team, answer the following prompts after implementing the lesson designed from the Mathematical Practices lesson-planning tool.

1. What level of student engagement with the lesson tasks did you observe? Describe how more direction, support, or scaffolding might have been provided if necessary.

2. Did students produce a variety of solutions or use a variety of strategies? If not, how might the structure of the tasks be redesigned to make them more open?

3. Describe any unexpected or novel solutions that would be worth remembering and incorporating into future classroom discourse.

4. What student misconceptions became evident during the lesson?

5. Which solutions were discussed and in what order? How did those choices support the classroom discourse? Upon reflection, might other choices have been more productive?

Figure 2.17: Reflection and debriefing prompts to follow implementation of the Mathematical Practices lesson-planning tool.

Visit **go.solution-tree.com/mathematicsatwork** to download a reproducible version of this figure.

Teachers should add lesson reflections, including sample student solutions, to each lesson plan for the unit and keep them on file. Expert teachers have the ability to explain complex ideas with clarity. Used effectively, the teacher learning process during a lesson study reflection benefits student and teacher understanding. Thus, an electronic sharing folder such as Google Docs, Dropbox, or the district's learning management system or repository makes these lesson notes readily available for lesson planning in following units and years. While this approach falls short of formal lesson study, such reflection provides ongoing opportunities for collaborative teams to continuously improve each lesson and learn from experience. Ultimately, this activity should increase the quality of team instruction and student learning. If your teams have not yet established such a repository, doing so is an important component of their collaborative planning. You should both expect and support the creation of an electronic team file as experts on your teams and from your mathematics instructional coaching support share their knowledge.

Team Progress

It is helpful for teams to diagnose their current reality and action during the unit. Ask each team to individually assess its progress on the eighth high-leverage team action using the status check tool in table 2.3. Discuss your perception of your teams' progress on using a lesson-design process for lesson planning and collective team inquiry. The real value in collaborating occurs in your discussions after they have tried the lesson. As your teams seek stage IV—sustaining—you increase the probability that all lessons contain an appropriate balance of higher- and lower-level-cognitive-demand mathematical tasks and provide opportunities for all students to benefit from teachers' formative assessment lesson planning.

Table 2.3: During-the-Unit Status Check Tool for HLTA 8—Using a Lesson-Design Process for Lesson Planning and Collective Team Inquiry

Directions: Discuss your perception of your team's progress on the eighth high-leverage team action—using a lesson-design process for lesson planning and collective team inquiry. Defend your reasoning.			
Stage I: Pre-Initiating	**Stage II: Initiating**	**Stage III: Developing**	**Stage IV: Sustaining**
We do not use a lesson-planning tool.	We plan for instruction using the Mathematical Practices lesson-planning tool or other lesson templates independently.	We develop common lessons, either using the Mathematical Practices lesson-planning tool or other lesson templates but do not discuss the implementation.	We develop and implement common lessons at least once per unit, either using the Mathematical Practices lesson-planning tool or other lesson templates.
We do not know if our lessons provide for student demonstrations of understanding.	We discuss student demonstrations of understanding but do not have a common agreement on how to achieve them.	We collaboratively agree on how students should demonstrate understanding, but we do not make instructional adjustments based on those agreements.	We ensure all lessons contain successful opportunities for students to demonstrate understanding.
We do not know about lesson study.	We have read about lesson study but do not create the time to do it.	We have engaged in a team lesson study but not as an ongoing practice.	We actively engage in a team lesson study once per unit and debrief in order to learn more about our students and to learn from each other.

Visit **go.solution-tree.com/mathematicsatwork** to download a reproducible version of this table.

Setting Your During-the-Unit Priorities for Team Action

When your school functions as a PLC, your grade-level and course-level collaborative teams must make a commitment to pursue the three high-leverage team actions outlined in this chapter.

> HLTA 6. Using higher-level-cognitive-demand mathematical tasks effectively
>
> HLTA 7. Using in-class formative assessment processes effectively
>
> HLTA 8. Using a lesson-design process for lesson planning and collective team inquiry

Ask each team, at least twice per year, to reflect together on the stages it identified with for each of the three team actions. Based on the results, ask each team to identify a priority for this aspect of its work together. Team members can use figure 2.18 (page 100) to focus time and energy on actions that are most urgent in their team preparation during the unit.

Remember, teams should always lesson plan from the students' point of view. Questions at the heart of the during-the-unit work within each team are, What will our students be doing during every aspect of the lesson? How will we provide formative feedback to them? How will they be expected to take action on that feedback? You can help your teams to focus on these few but complex tasks and make them matter at a deep level of implementation.

Our attention now turns to chapter 3, and to steps four and five of the PLC teaching-assessing-learning cycle, with a focus on implementing a formative assessment process in response to the end-of-unit common assessments, your teams' and students' formative response to the data revealed at the end of the unit, and the impact on instruction during the next unit of study.

Directions: Identify the stage you rated your team for each of the three high-leverage team actions, and provide a brief rationale.

6. Using higher-level-cognitive-demand mathematical tasks effectively

    Stage I: Pre-Initiating   Stage II: Initiating   Stage III: Developing   Stage IV: Sustaining

    Reason: _____

    _____

    _____

    _____

    _____

7. Using in-class formative assessment processes effectively

    Stage I: Pre-Initiating   Stage II: Initiating   Stage III: Developing   Stage IV: Sustaining

    Reason: _____

    _____

    _____

    _____

    _____

8. Using a lesson-design process for lesson planning and collective team inquiry

    Stage I: Pre-Initiating   Stage II: Initiating   Stage III: Developing   Stage IV: Sustaining

    Reason: _____

    _____

    _____

    _____

    _____

With your collaborative team, respond to the red light, yellow light, and green light prompts for the high-leverage team actions that you and your team believe are most urgent.

Red light: Indicate one activity you will stop doing that limits effective implementation of each high-leverage team action.

Yellow light: Indicate one activity you will continue to do to be effective for each high-leverage team action.

Green light: Indicate one activity you will begin to do immediately to become more effective with each high-leverage team action.

Figure 2.18: Setting your collaborative team's during-the-unit priorities.

Visit **go.solution-tree.com/mathematicsatwork** to download a reproducible version of this figure.

CHAPTER 3

After the Unit

You can't learn without feedback. . . . It's not teaching that causes learning. It's the attempts by the learner to perform that cause learning, dependent upon the quality of the feedback and opportunities to use it. A single test of anything is, therefore, an incomplete assessment. We need to know whether the student can use the feedback from the results.

—Grant Wiggins

Teachers have just taught the unit and given the common end-of-unit assessment (developed through high-leverage team actions 3 and 4). What should happen next? Did the students reach the proficiency targets for the essential learning standards of the unit? As a school leader, how do you know? More important, what are the responsibilities for each of your collaborative teams after the unit ends?

The after-the-unit high-leverage team actions support steps four and five of the PLC teaching-assessing-learning cycle (see figure 3.1, page 102).

Think about when your teachers pass back an end-of-unit assessment to their students. Did assigning the students a score or grade motivate them to continue to learn and to use the results as part of a formative learning process? Did the process of learning the essential standards from the previous unit stop for the students as the next unit began? In a PLC culture, the process of student growth and demonstrations of learning *never* stop.

Recall that in step one of the cycle (chapter 1), your collaborative teams unpacked and paced the essential learning standards for the unit, identified meaningful mathematical tasks to balance cognitive demand, created the common assessment instruments for the unit with scoring rubrics, agreed to proficiency targets for the essential learning standards, and agreed to common homework assignments that support students' conceptual understanding and procedural fluency.

In steps two and three of the cycle (chapter 2), the focus was on what students were doing in class—how your collaborative teams developed and implemented lessons that encouraged students to engage in and demonstrate the Standards for Mathematical Practice and incorporated the effective use of in-class formative assessment processes. Your teams used lesson-planning tools to organize the mathematical tasks your teams developed in chapter 1 into class-ready lessons for collective inquiry.

Now, after the unit has ended, is a moment of transition. Teachers must score their assessments, and begin the next unit. Everything needs to be ready. In steps four and five of the PLC teaching-assessing-learning cycle, your teachers (with your help) design a process to accurately score the common assessment instruments. They also examine the role students play in using teacher feedback to improve and take action on their learning progress.

Also, as the unit ends, teachers reflect on student performance, adjust their decisions about the next unit based on that information, and take action as needed based on the assessment results (step five of the PLC teaching-assessing-learning cycle and HLTA 10).

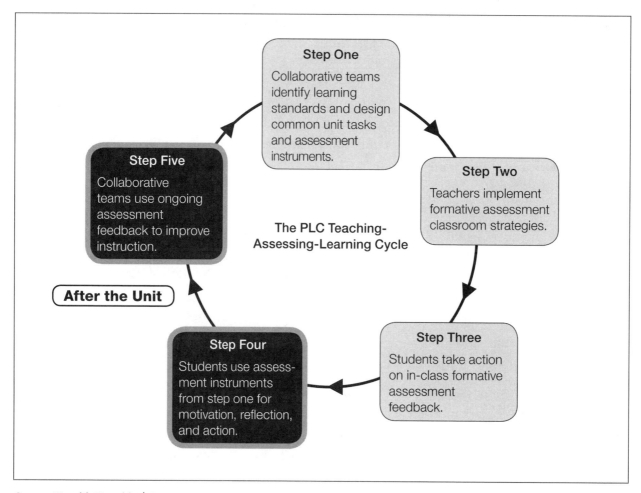

Source: Kanold, Kanold, & Larson, 2012.

Figure 3.1: Steps four and five of the PLC teaching-assessing-learning cycle.

To complete the analysis expected in steps four and five of the teaching-assessing-learning cycle, your collaborative teams engage in the final two high-leverage team actions, both of them representing part of a formative assessment process.

> HLTA 9.   Ensuring evidence-based student goal setting and action for the next unit of study
>
> HLTA 10. Ensuring evidence-based adult goal setting and action for the next unit of study

As John Hattie (2012) states:

> My role as a teacher is to evaluate the effect I have on my students. . . . This requires that teachers gather defensible and dependable evidence from many sources, and hold collaborative discussions with colleagues and students about this evidence, thus making the effect of their teaching visible to themselves and to others. (p. 19)

The PLC teaching-assessing-learning cycle allows teacher and student reflection around evidence of learning on a unit-by-unit basis throughout the year and provides for a progression of learning beyond the end of the unit. Testing is not an end; it is a means to advance student and adult learning. It is your leadership that will provide this cultural mindset as a strategic narrative in your school or district.

HLTA 9: Ensuring Evidence-Based Student Goal Setting and Action for the Next Unit of Study

Indeed, the whole purpose of feedback should be to increase the extent to which students become owners of their own learning.

—Dylan Wiliam

Think about the last end-of-unit assessment your teachers graded (scored) and returned to your students. What was the typical student response to the unit assessment feedback? Did students in your school or district use this feedback to further their understanding of the essential learning standards? Collaborative teams that engage students in the teaching-assessing-learning cycle use evidence of student learning to re-engage students in learning in step four to ensure the end-of-unit assessment grade or score does not stop student learning. Wiliam (2011) states it like this: "As soon as students get a grade on a test, the learning stops" (p. 123). Learning should not stop when teachers pass back tests that include meaningful feedback. Remember that delivering a score is the least-important aspect of returning the assessment. Does your school culture *require* student reflection and action on assessment performance?

Wiliam (2007) makes the distinction between using assessment instruments for the purposes of (1) monitoring, (2) diagnosing, or (3) formatively assessing. He states:

> An assessment monitors learning to the extent it provides information about whether the student, class or school is learning or not; it is *diagnostic* to the extent it provides information about what is going wrong, and it is *formative* to the extent it provides information about what [students are] to do about it. (p. 1062)

Thus, a key formative assessment feature after the unit ends will be the schoolwide Tier 2 process you have in place for students to use and to respond to their assessment results.

Once again, recall there are four critical questions every collaborative team in a PLC asks and answers on an ongoing, unit-by-unit basis.

1. What do we want all students to know and be able to do? (The essential learning standards)

2. How will we know if they know it? (The assessment instruments and tasks teams use)

3. How will we respond if they don't know it? (Formative assessment processes for intervention)

4. How will we respond if they do know it? (Formative assessment processes for extension and enrichment)

High-leverage team action 9—ensuring evidence-based student goal setting and action for the next unit of study—guarantees your students use common end-of-unit assessment results as part of a formative assessment process that responds to critical questions three and four. Using evidence of student learning to guide instruction is so fundamental that eliciting and using "evidence of student thinking" is one of only eight research-informed instructional strategies outlined in *Principles to Actions* (NCTM, 2014, p. 53). What is the expected after-the-unit teacher, teacher team, and *school* response if students demonstrate there are certain standards they do or do not know?

High-Leverage Team Action	1. What do we want all students to know and be able to do?	2. How will we know if they know it?	3. How will we respond if they don't know it?	4. How will we respond if they do know it?
After-the-Unit Action				
HLTA 9. Ensuring evidence-based student goal setting and action for the next unit of study			■	■

■ = Fully addressed with high-leverage team action

The What

Real-Life Leadership Scenario

We were asked to consult with a middle school that was not showing improvement on their state mathematics assessment, despite the fact they had a built-in intervention time for students who were struggling in math. The school had created a math intervention course at each grade level to support students who might struggle in the curriculum. Students identified as at risk in mathematics— significantly below grade level—were enrolled in both their grade-level math course and a math intervention class, receiving two periods of math instruction. In analyzing why the intervention wasn't having the desired results, it became clear that although the intervention did provide support, the intervention was not fluid in its response to student learning needs that emerged during the year. Students were only removed or added to the intervention support at the end of a grading period. Additionally, the intervention was too disconnected from the learning experiences in the on-level mathematics course. The intervention was not sufficiently integrated with students' regular instruction, was not based on learning the essential standards of the grade level, and was not flexible enough to meet student needs in a timely manner.

So we recommended to the principal a switch to a more fluid math intervention structure based on learning standard student performance on the teacher teams' common unit assessments. With this approach, collaborative teams began to analyze assessment results together and then regroup students into three distinct one-week interventions for targeted support based on student proficiency for a specific learning standard.

Some students might receive one week of targeted support, others two, three, or none after each assessment cycle. The targeted support became fluid, students moved in and out as needs arose, and support became linked directly to the grade-level curriculum learning standards. In order to receive the targeted support, the student was moved during the normal daily schedule. This fluid approach to math intervention resulted in gains on the state math assessment at each grade level that were at least 50 percent greater than the gains for students enrolled in the traditional static math intervention. Based on its effectiveness, other middle schools in the district converted toward a more fluid approach to math intervention.

As a school leader, high-leverage team action 9 requires your direct help and support. In many K–12 classrooms, students receive their assessment back from the teacher, look at the score, and place the paper in their notebook or backpack. This action of locking away the unit assessment (and moving on) wastes a valuable formative learning opportunity for students.

By working together on this high-leverage team action with your teacher teams, you will begin to reduce the variance that often occurs in student learning across classrooms. The students in your school begin to reflect on successes and also focus their next-step goals and actions based on evidence of areas of weakness (standards not yet learned) after the unit of study. All students in your school should see assessment results as a means to better understand their current mathematical knowledge and be able to use the information on the assessment to improve their mathematical understanding. Thus, assessment becomes an integral part of the learning process for students. It is why assessment is embedded in the middle of the PLC teaching-assessing-learning cycle from unit to unit.

Using assessment results effectively requires two very important components. First, students need to be given FAST (fair, accurate, specific, and timely) feedback on their end-of-unit assessment work in order to improve their mathematical understanding. Second, students must receive the opportunity to learn how to use their assessment results in formative ways in order to take timely action on teacher feedback.

The How

As a first step, brainstorm with your collaborative teams ways they might include students in a formative assessment process that requires response to teacher feedback followed by student action on previous units' results. Make a list of your teams' ideas, and return to them as you continue through this chapter.

Understanding the Characteristics of Effective Feedback: FAST—Fair, Accurate, Specific, Timely

Step four (figure 3.1, page 102) begins with the end-of-unit common assessment feedback teachers provide to students. Recall from the in-class formative assessment work with HLTA 7 (page 75) that the essential characteristics of effective in-class feedback were that feedback is fair, accurate, specific, and timely. These same components of effective feedback hold true once again as teachers pass the tests back to students.

1. **Feedback must be Fair:** Effective feedback on the test rests solely on the quality of the work the student demonstrates, not on other characteristics of the student.

2. **Feedback must be Accurate:** Effective feedback on the test acknowledges what students are currently doing well and correctly identifies errors they are making. According to Stephen Chappuis and Rick Stiggins (2002), "Effective feedback describes why an answer is right or wrong in specific terms that students understand" (p. 42).

3. **Feedback must be Specific:** Teachers' test notes and feedback "should be about the particular qualities of [the student's] work, with advice on what he or she can do to improve, and should avoid comparison with other pupils" (Black & Wiliam, 2001, p. 6). Help each teacher try to find the right balance between being specific enough that the student can quickly identify the error or logic in his or her reasoning but not so specific that the teacher does the correction work for him or her. Does test feedback help *students* correct their thinking as needed?

4. **Feedback must be Timely:** Effective feedback on the end-of-unit assessment must be provided in time for students to take formative learning action on the results before too much of the next unit has taken place. As a general rule, teachers should pass back end-of-unit assessments and the results for proficiency to students within forty-eight hours of the assessment.

> To begin your teams' development of using feedback effectively, have your teams complete the questions in figure 3.2 during a team meeting. Be sure to review the teams' responses and provide them with your feedback on the tool.

Increasing Accuracy and Decreasing Variance of Student Scoring and Feedback

There will be variability in how team members score similar items unless team members agree on the scoring of the mathematical tasks on the test and the accuracy of the feedback they give to students. If they do not agree about scoring, grading inequity will occur within the grade level or course, leading to an unreliable scoring process from teacher to teacher. Recall that this team discussion about how to score the exam was to have taken place before the unit began (see HLTA 4, p. 41). However, if not, then it needs to happen before the teachers start their grading of the actual unit assessment.

As an example, consider the photography task and sample student response in figure 3.3 (page 108). How would you expect middle school teachers to score this seventh-grade mathematics task based on the student work? The teacher team decided this task was worth six points. What score could the student receive for his or her response: 2 out of 6? 3 out of 6? 5 out of 6? Unless the teachers calibrate their scoring together on some student papers, individual students will receive a very different score or grade on this assessment task. Remember, accuracy of scoring across all teachers on the team is the goal.

Directions: Use your responses to these questions to guide the feedback you would offer students based on their responses on the unit assessment.

1. How did the strategies the student used demonstrate his or her understanding of the essential learning standard the task assesses?

2. How did the feedback you generated build on the student's strengths to address the learning needs for the task?

3. How did the feedback you generated guide the student to understand an error (if you believe there was an error) without being too directive?

4. How did the feedback you generated recognize student effort and accuracy?

5. How did the feedback recognize the Mathematical Practices the student used for the task?

6. In what ways did you provide the feedback in enough time to allow for effective student action?

Figure 3.2: Collaborative team student feedback discussion tool.

Visit **go.solution-tree.com/mathematicsatwork** to download a reproducible version of this figure.

7. Photographers Ryan and Alex get paid per photography session. Ryan is paid a set-up fee of $25 plus $10 per hour. Alex is paid a set-up fee of $20 plus $11 per hour.

   a. Write an expression for each photographer. Use *h* to represent hours worked.

   Ryan: $\underline{\$25 + \$10h}$ Alex: $\underline{\$20 + \$11(h)}$

   b. Mr. Herrera wants to take pictures of his new baby boy. If the photo session will take **2 hours**, who should he choose to take pictures? **Explain your reasoning**.

   $25 + 10(2)$ $20 + 11(2)$ =   *Alex because it costs less money. The flat fee is less and the per hour is only a dollar more.*
   $\$45$ $\$42$

   c. If Mr. Herrera wants to take pictures for 5 hours, should he choose Ryan or Alex as his photographer? Explain your reasoning.

   $\begin{array}{r} 25 \\ +\ 50 \\ \hline 75 \end{array}$ $25 + 10(5)$ $20 + 11(5)$ $\begin{array}{r} 11 \\ \times\ 5 \\ \hline 55 \\ +\ 20 \\ \hline 75 \end{array}$
                          $\$75$ $\$75$

   Either photographer because both have the same price for 5 hrs.

Source: Adapted with permission from Aptakisic Junior High School, Buffalo Grove, Illinois.

Figure 3.3: Sample student response for photography task.

Expect your teachers to select their most recent common end-of-unit assessment. Choose two or three sample student tests from each teacher's assessment. Ask teachers to work as partners or as a complete team to score the test together. Then, they should share the scores and the nature of the feedback they offered the students. This will help you to gain more confidence that your teachers' grading is accurate. In the event there is a wide discrepancy in the student score, your teachers can either ask a third teacher to score and negotiate a common understanding, or, the teacher partners rescore until they are more calibrated with one another.

Be sure to also discuss the timeliness of teacher feedback to students as well. Do you monitor for this? What happens if you have a teacher who does not pass back an end-of-unit test until one to two weeks into the next unit? Accurate and specific feedback will not help students if it is not also *timely*. If teachers provide feedback to students well after giving the assessment, the feedback will have minimal impact.

Ask your teachers to show you how they are providing *immediate feedback* to students on the end-of-unit assessment. Is it timely and effective for students to take action on the results?

> Do you know when one mathematics unit ends and a new mathematics unit begins? This is a very important moment in your school or district for you to monitor, as it is at this very moment that intervention must kick in on the previous unit's standards while at the same time the next mathematics unit begins. Your teacher teams will need time and support to transition successfully. They need your help.

Requiring Student Action on Common End-of-Unit Assessment Feedback

Think of this expectation for student action as part of a Tier 2 response within the RTI model discussed in chapter 2 (page 85). Tier 2 represents a targeted response to student learning and is part of your effort to answer the third critical question of a PLC culture, *How will we respond if they don't know it?* For an in-depth discussion of RTI and Common Core mathematics, see chapter 5 from our 2012 *Common Core Mathematics in a PLC at Work* series (Kanold, Kanold, & Larson, 2012).

How can the students in your school become active partners in this formative end-of-unit process? Your teams can build time during the instructional day to ensure the following four elements. Students:

> 1. Examine their end-of-unit results to determine whether or not they met the proficiency target for the unit's essential learning standards
>
> 2. Address errors on the assessment and use the assessment results to set goals and form a plan (either collaboratively between teacher and student or using a teacher-supplied plan)
>
> 3. Act on their plan during the next unit
>
> 4. Act on opportunities to improve their score on the end-of-unit assessment
>
> Note: Use these protocols to examine the quality of the current response in your school.

Simply assigning a grade to an assessment is least useful in motivating students to further their understanding (Kanold, Kanold, & Larson, 2012). Students must use the feedback from the unit assessment as a formative learning opportunity, instead of a "test once, then done" approach. Teacher feedback will allow students to reflect on their strengths and challenges from the assessment. When teachers return the scored assessment to the class, they need to plan for student time to self-assess results and make a plan of action to retool their learning. Wiliam (2011) describes that effective feedback should:

> Cause thinking: It should be focused; it should be related to the learning goals that have been shared with students; and it should be more work for the recipients than the donor. Indeed, the whole purpose of feedback should be to increase the extent to which students are owners of their learning. (p. 132)

Figure 3.4 (page 110) shows one type of student action plan model you could use with students as your teams return the results of an end-of-unit assessment. Notice the use of student-friendly language and sentence starters to encourage student participation at all grade levels. Please be aware that teachers will need to provide additional support for young learners and emerging readers.

Directions: How did you do? What do you need to do now? Think about this and answer these questions.

Student Name: **Teacher:**

Unit:

My Essential Learning Standards:

What were my learning standards for this unit?

"I can . . . _____."

"I can . . . _____."

"I can . . . _____."

My Learning Performance:

Am I getting closer to learning each standard? Did I meet the learning target?

"I can . . . _____."   Yes   No

"I can . . . _____."   Yes   No

"I can . . . _____."   Yes   No

For any learning standards you answered no:

I got _____ on this test. I will get _____ next time.

My Actions:

What will I do to get closer to my learning targets next time?

_____ Ask my teacher for help.

_____ Ask my partner for help.

_____ Ask for more time.

_____ Do more problems on my own.

_____ Get help online.

_____ Other actions: _____

_____

_____

_____

_____

_____

_____

_____

_____

Figure 3.4: Student action plan template for response to an end-of-unit assessment.

Visit **go.solution-tree.com/mathematicsatwork** to download a reproducible version of this figure.

You also need to help your teams by providing the structure and time necessary so students can take the actions they determine during the self-assessment. The best way to do this is to ensure that every day there is a period of time designated for this purpose—for providing extra time and support for the essential learning standards of the previous unit. You can go to the www.allthingsplc.info website for ideas on how to provide fluid intervention time around support for learning standards. During this time, students can use figure 3.5 (page 112) as a helpful tool to review their tests and self-assess.

Any student who does not meet or exceed the proficiency targets for the previous unit must receive extra support from your school and from the collaborative team. This is a great opportunity for you to join forces with all team members and other school support personnel to provide intervention, extra practice on key tasks, and enrichment for those students meeting all essential learning standards.

Every student should be involved in some type of response to his or her performance from the previous unit. The results of the end-of-unit assessment should drive the intervention and support for your students and have an element of student goal setting as well (Kanold, Kanold, & Larson, 2012).

Using Re-Engagement Versus Reteaching

To facilitate this response on the part of students, each of your collaborative teams will need to create re-engagement strategies for students that are aligned to the essential learning standards. What would re-engagement look like after the unit of instruction? Keep in mind that re-engagement is different than reteaching.

Reteaching implies teachers present the content the same way as the first learning experience. Your leadership role is to ensure that your teachers avoid reteaching. When reteaching standards, teachers often do not use student misconceptions from the unit assessments to engage thinking about the concepts in a new way. *Re-engagement* is a system of targeted intervention that utilizes data regarding student misconceptions to engage students differently in mathematical concepts. Effective collaborative teams review student misconceptions and craft re-engagement lessons, intervention strategies, and tools that are directly tied to the learning standards on the assessment (Foster, 2008).

Thus, your teams set up processes that allow students to:

1. Reflect on the feedback from the unit assessment (usually when you pass back tests in class)

2. Use the specific feedback for reflection, motivation, and action (re-engagement)

3. Receive an improved score based on the assessment results (grades reflect actual learning of the standards)

As students reflect on the current reality of their understanding throughout the unit, they create a plan to take action on standards they do not understand *prior* to the next end-of-unit assessment. Students cannot choose to not re-engage in learning. This is not an option. Your teams must ensure that Tier 2 RTI re-engagement structures for students are timely.

The intent of student self-assessment of end-of-unit performance is to build student responsibility for learning as part of a formative learning process and a schoolwide Tier 2 RTI response.

Directions: Review your test and complete the following.
Name:
Date: **Class:**

1. I did well on these standards . . .

2. What does the test tell me about what I do well?

3. I did not do well on these standards . . .

4. What does the test tell me about what I still need to learn?

5. Did all of my answers make sense?

6. Did I show all of my work and check my answers for each problem?

7. I am still confused about . . .

Which standards do I still need to learn?	What will I do to learn them?
What did I learn this time that I will apply to future mathematics tests?	

Source: Adapted from Wallace, 2013.

Figure 3.5: Tool for student self-assessment and feedback.

Visit **go.solution-tree.com/mathematicsatwork** to download a reproducible version of this figure.

Part of your role as a leader is to decide how to broker and create the time and the space needed for teachers to implement a systemic Tier 2 intervention at the end of every unit. Although the answer to the time question is dependent on the circumstances at your school or district, intervention should be based on a few basic protocols:

1. It should be fluid. (Students can move in and out of the intervention once they demonstrate proficiency on the standards.)

2. It should be directly connected to the mathematics learning standards for the previous unit and the current unit the student is learning. (It should not be a random intervention of math standard support.)

3. It should be owned and shared by the teacher team. (This increases the likelihood that it is connected to the daily teaching for the students.)

4. It should be required. That is, any student demonstrating a lack of proficiency on a standard should be required to participate in mathematics intervention and support.

Student-to-student feedback is a vital component of the entire educative process, and the goal-setting process at the end of each unit (for the next unit) is part of that student collaboration. This collaboration with peers and ownership of the learning pathway will engage students more deeply in the learning process and provide evidence for each student that effective effort is the route to mathematical understanding and success. Visit **go.solution-tree.com/mathematicsatwork** for more details and examples of this process.

Team Progress

It is helpful for your teams to diagnose their current reality and action after launching the unit. Ask each team to individually assess its progress on the ninth high-leverage team action using the status check tool in table 3.1 (page 114). Discuss your perception of your teams' progress on ensuring evidence-based student goal setting and action for the next unit of study. It matters less which stage your teams are at and more that team members are committed to working together and understanding how to support students who still need help after the end-of-unit assessment as teams seek stage IV—sustaining.

Your teams' careful, honest, and open answers on the status check tool will allow for movement toward effective implementation of formative assessment strategies to assist students and teachers alike in the learning process. After reflecting on some of the work you and your teams have done in chapters 1 and 2, you must now turn to the final high-leverage team action, HLTA 10—ensuring evidence-based adult goal setting and action for the next unit of study.

Table 3.1: After-the-Unit-Ends Status Check Tool for HLTA 9—Ensuring Evidence-Based Student Goal Setting and Action for the Next Unit of Study

Directions: Discuss your perception of your team's progress on the ninth high-leverage team action—ensuring evidence-based student goal setting and action for the next unit of study. Defend your reasoning.			
Stage I: Pre-Initiating	**Stage II: Initiating**	**Stage III: Developing**	**Stage IV: Sustaining**
We do not discuss whether our test feedback is fair, accurate, specific, or timely.	We discuss how our test feedback should be fair, accurate, specific, and timely, but we do not know what other team members actually do.	We provide fair, accurate, specific, and timely feedback to students but do not discuss its impact as a collaborative team.	We provide fair, accurate, specific, and timely feedback to students, and we discuss the impact of this feedback as a collaborative team.
We do not provide students with opportunities to respond to the feedback from the end-of-unit assessment.	We provide some constructive feedback to students on the end-of-unit assessment, but we do not require them to respond to the feedback.	We require students to correct their errors on the end-of-unit assessment.	We require students to correct their errors and identify the learning standards that are strengths and weaknesses.
We do not know the type of end-of-unit assessment feedback other team members use.	We do not have a team process in place for student response to the end-of-unit assessment results.	We work with each student to identify a plan for improvement and action based on end-of-unit results and improvement.	We work with each student to complete and carry out a plan for improvement and action based on end-of-unit results.

Visit **go.solution-tree.com/mathematicsatwork** to download a reproducible version of this table.

HLTA 10: Ensuring Evidence-Based Adult Goal Setting and Action for the Next Unit of Study

[High-impact teaching] requires that teachers gather defensible and dependable evidence from many sources, and hold collaborative discussions with colleagues and students about this evidence, thus making the effect of their teaching visible to themselves and to others.

—John Hattie

The final high-leverage team action and second after-the-unit pursuit is to consider how your teams use the results of the end-of-unit assessment for their own reflection and learning. While it is important for students to use feedback to monitor and improve their understanding of mathematics, it is just as important for your teachers to reflect on the unit assessment data *as a team.*

Your teams need to use the unit's assessment instrument results to monitor and evaluate their instruction. This provides a formative assessment learning process for your teams, not just for the students. This high-leverage team action ensures your teams reach clarity on how to effectively respond *after* the unit ends, addressing the third and fourth PLC critical questions: How will we (the collaborative team) respond if they don't know it? How will we respond if they do know it?

You will need to help your teams find the time for this action as they complete each mathematics unit in the curriculum during the school year.

High-Leverage Team Action	1. What do we want all students to know and be able to do?	2. How will we know if they know it?	3. How will we respond if they don't know it?	4. How will we respond if they do know it?
After-the-Unit Action				
HLTA 10. Ensuring evidence-based adult goal setting and action for the next unit of study			▢	▢

▢ = Fully addressed with high-leverage team action

The What

──────────── **Real-Life Leadership Scenario** ────────────

I [Tim] was in a meeting with a grades 3–5 vertical team of teachers and their principal. I asked them what seemed at the time as a bit of an odd question, "How do you know your hard work mattered?" In other words, how did they know they were "winning" each day, each week, and in each unit? They were not sure. We brainstormed a bit around what does it mean to be successful on a weekly basis in their mathematics classes? Their answers were a bit vague, but ultimately they settled on: "We will know next May when the students take the state test." That seemed a little too late to me. They need to know this week.

I then asked the team members to list the evidence they could show me so far (it was October) that they were having a successful school year. They asked why I wanted to know. I told them a simple answer: cake. How did they know to celebrate short-term wins and have cake together if they never reflected on the outcomes of their work at the end of every unit and then use those outcomes to impact their own instruction in the next unit? Cake eventually became an inside joke ("Let them eat cake!") and a steady part of team meetings.

Teams can consider the following questions when collecting evidence of effectiveness from common end-of-unit assessments.

- What went well in the unit?
- How well did students understand the essential learning standards of the unit?
- Which students need additional time and support to become proficient?
- Which students would benefit from an extension of the standard due to demonstrated proficiency?
- How well did we provide feedback to the students during the unit?
- How did the results vary by teacher? What are areas that warrant attention and need improvement for individual teachers as well as the team?
- How will we respond to the evidence of learning standards that did not result in student success?

The answers to these questions should be based on personal subjective reflection and supported by the data from the common end-of-unit assessment. At this moment—step five in the PLC teaching-assessing-learning cycle (figure 3.1, page 102)—you charge your teams with the task of providing you with a brief and timely reflection on how their efforts during the unit did or did not meet with success. (The team leader can provide this report.)

Note that it is important to do this reflection before teams reach too deeply into the next unit of instruction if the results are to have an impact on their instruction in the next unit. You will need to keep track of each team's unit calendars, know when the units end throughout the year, and make sure your teams have the time to meet to get this work done.

From a practical point of view, this sets up a cycle of teacher planning and reflection about ten times per year. This may seem too often at first, but with practice, it becomes the norm of team practice in a PLC culture. A key feature of the effective use of common end-of-unit assessment is how well your teams reflect on student acquisition of the essential learning standards for the unit.

The How

Ask your teams to examine the most recent end-of-unit common assessment they have implemented. Consider the results of the end-of-unit assessment and what questions the data brought to mind. Ask your teachers to respond to all of the questions in figure 3.6 based on an objective review of the results.

Directions: Answer each of the following end-of-unit assessment questions with your collaborative team.

Team: _____   **Teacher:** _____

1. What went well overall in the unit?

2. What do the common end-of-unit data reveal about student performance on the essential learning standards with which students did well?

3. How did each teacher achieve these student successes?

4. What do the common end-of-unit data reveal about student performance on the essential learning standards with which students did not do well?

5. How well did we provide feedback to our students during the unit on the learning standards with which students struggled?

6. What elements of the unit instruction will we need to improve for future use?

7. What impact do the results from this unit have on instruction during our next unit based on our team's reflections? (RTI Tier 1)

8. How will our team provide for student re-engagement around essential learning standards that need more time, support, and focus for student learning? (RTI Tier 2 targeted team response)

Figure 3.6: Collaborative team data-analysis protocol.

Visit **go.solution-tree.com/mathematicsatwork** to download a reproducible version of this figure.

> You support the important work of HLTA 10 by expecting your teacher teams to complete and send the data-analysis protocol tool to you within a timely manner at the end of every mathematics unit. You need to then read and review their responses on the tool and then provide them with feedback as needed.

As your teachers respond to the questions in figure 3.6, make sure they honor a bias-free environment. Often, when looking at data, teachers make inferences beyond the scope of the data—perhaps about the effect of the learning environment on certain students, how the field trip the previous day impacted the results, or how some students have support at home that others do not receive. While these are all potential factors in determining results, at this stage, collaborative teams should focus on how the data reflect *the efforts of the team* during instruction of the unit. You may want to collect this form from team leaders at the end of every unit.

Scoring the Common Assessments Collaboratively

It is a great opportunity for learning when you help your teachers understand how to score assessments together. A collaborative analysis of the common student work will also provide insight into collaborative team members' mathematical thinking and support the consideration of multiple ways that students and teachers represent their thinking. You can use figure 3.7 (page 119–120) to facilitate this team activity.

Help your teachers to develop a consensus score for each sample student response to the chosen mathematical tasks. Discussions based on what the student work represents will focus the team on ensuring that the score is based solely on the mathematical understanding and appropriate problem-solving strategies. It is important for the team to score fairly and consistently, so practicing common scoring and analysis of the same items should occur often within collaborative teams.

You may need to broker a special schedule or specific time during the normal work day to allow your teacher teams time to be able to do the analysis required at the end of a unit. If this does not occur during normal team time, then ask each team to submit a plan to you as to how they will complete the required analysis of the student results from the unit assessment.

Figure 3.8 (page 121) shows an example of a learning standard tracking tool that some teacher teams can use for more detailed data collection after the end of each assessment. The example in figure 3.8 is from a sixth-grade team on a unit 1 assessment. Teachers gathered the data evidence individually and brought their completed form and student work to the collaboration team time immediately following the unit assessment.

In the example, the team chose to track the number of students proficient on each essential standard (of various grain size) to identify strengths within the unit and then plan re-engagement strategies for concepts that students consistently missed. In areas with minor misconceptions, collaborative teams planned upcoming lessons and used scaffolding to address common misconceptions.

Specific strategies for the next unit could include using warm-up tasks that cover both the prerequisite knowledge for the lesson and previous unit misconceptions or adding review items to current homework to encourage continual practice on concepts that students mastered during the unit.

Directions: Collect sample student work from your most recent common end-of-unit assessment. Make copies so that each team member has a complete set of student test samples to score, removing student names and numbering them for easy reference. Each team member should score the examples individually and make notes justifying his or her scoring decisions. Next, share your scoring and notes.

1. Did everyone score the student work for each mathematics task the same?

2. What reasons do team members have for their scores for the student work being different?

3. Develop a consensus score for each sample item. Ensure the score is based on mathematical understandings and appropriate problem-solving strategies as demonstrated by the student.

4. What strategies did you use to agree as a team on a common score for each task on the test?

5. Organize your reflections on the students' mathematical thinking. What are the patterns of thinking most prevalent for the tasks? What are the misconceptions that stood out? Did students use strategies on the test questions that were surprising?

Figure 3.7: Collaborative team scoring and analysis activity.

continued →

Student Name	Description of Student Work	Student Strengths	Students Misconceptions	Future Work With Student

Implications for instruction:

Figure 3.7: Collaborative team scoring and analysis activity (continued).

Visit **go.solution-tree.com/mathematicsatwork** to download a reproducible version of this figure.

Unit 1: Add, Subtract, Multiply, and Divide Rational Numbers		Teacher			
		1	2	3	4
Content standard cluster: Apply and extend previous understandings of multiplication and division to divide fractions by fractions.		61%	45%	67%	70%
6.NS.1	Interpret and compute quotients of fractions, and solve word problems involving division of fractions by fractions.	61%	45%	67%	70%
Content standard cluster: Compute fluently with multidigit numbers and find common factors and multiples.		78%	61%	69%	79%
6.NS.2	Fluently divide multidigit numbers using the standard algorithm.	78%	79%	76%	89%
6.NS.3	Fluently add, subtract, multiply, and divide multidigit decimals using the standard algorithm for each operation.	77%	50%	63%	71%
Content standard cluster: Compute fluently with multi-digit numbers and find common factors and multiples.		77%	37%	37%	58%
6.NS.4	Find the greatest common factor of two whole numbers less than or equal to 100 and the least common multiple of two whole numbers less than or equal to 12. Use the distributive property to express a sum of two whole numbers 1–100 with a common factor as a multiple of a sum of two whole numbers with no common factor.	77%	37%	37%	58%
Content standard cluster: Solve real-world and mathematical problems involving area, surface area, and volume.		61%	38%	70%	75%
6.G.2	Apply the formulas $V = l\,w\,h$ and $V = b\,h$ to find volumes of right rectangular prisms with fractional edge lengths in the context of solving real-world and mathematical problems.	61%	38%	70%	75%

Source for standards: NGA & CCSSO, 2010, pp. 42, 44, 45.

Figure 3.8: Sample unit 1 learning standard tracking tool.

Visit **go.solution-tree.com/mathematicsatwork** to download a reproducible version of this figure.

Teachers should complete this analysis individually and then meet to review the results and dig deeper into the student work to analyze student misconceptions. This sixth-grade team identified two essential learning standards (6.NS.1 and 6.NS.3) that challenged students. In this case, the team members organized their analysis into bigger essential standard categories—around the content standard clusters for each essential standard. During the year, they measured student progress in the progression of these content standard clusters as they unfolded in each unit. Upon reviewing student work on the test, teachers were able to create re-engagement strategies to target student misconceptions in the next unit of instruction.

The team also noted that some teachers had strengths with respect to particular content. Teacher four crafted a re-engagement lesson for 6.G.2, teacher one crafted a lesson for 6.NS.3, and the team shared the work for 6.NS.2. Then, team members created a schedule for homeroom or intervention time (after school for some students) and posted the schedule in their classrooms so all students could attend the targeted Tier 2 intervention based on the standard that the individual students were still not proficient on, even though students might attend an intervention session with a teacher other than their regular classroom teacher.

Some of the best conversations for student understanding occur when teachers discuss and review student work and patterns of errors. Teachers are sometimes limited in their perspective on how to teach a concept. One advantage of the team approach is to open up teachers to new perspectives and to re-engage students in multiple solution strategies.

Setting Teacher Team Goals for Improvement

In order for your collaborative teams to develop a cycle of continuous improvement, they need to set and reflect on collaborative team achievement goals on a unit-by-unit basis. These goals have the same structure as the goals used with students. Consider the structure of clear and manageable team goals using the SMART acronym (O'Neill & Conzemius, 2006).

- **Strategic and specific:** Our goals are aligned to targeted essential learning standards.

- **Measureable:** We know when we achieve the essential learning standards.

- **Attainable:** We can achieve the goals through collaborative team efforts.

- **Results oriented:** We clearly define the outcome.

- **Time bound:** We have a specified time within which to achieve the goals.

Your teams should set SMART goals (proficiency goals) for each unit around the essential learning standards as they utilize the assessment results from one unit to prepare for the next unit. Your teams can work on developing manageable short-term and long-term goals as well as orchestrating celebrations, for example, when students reach their goals on the essential learning standards.

As an example, consider the data from the three essential learning standards (CCSS mathematics content standard clusters) for the sixth-grade unit in figure 3.8 (page 121). This team had set SMART goals of 75 percent proficiency for all students in the course. Team members should celebrate meeting this short-term goal for the essential learning standard—"Fluently divide multidigit numbers using the standard algorithm" (6.NS.2). However, they need to plan for student re-engagement in the other essential standards based on the results of the assessment data.

You can use the questions in figure 3.9 to help teachers analyze their team SMART goal performance and their plans for instruction moving forward. If your teams are new to the SMART goal process, visit www.allthingsplc.info for more insight.

In setting unit-by-unit student performance goals, each collaborative team needs to keep in mind that the focus should remain on student achievement—how can teachers help students grow in their mathematical understanding along all of the learning standards for the unit? Teams should use the student data from the assessments to consider the revision of both instructional and assessment practices.

As a final point, consider the following questions with your collaborative teams in regard to the formative assessment process and use of end-of-unit assessment instruments.

- Did the tasks balance cognitive demand and rigor? Did students demonstrate this in their responses?

- In what ways did students demonstrate proficiency with the Mathematical Practices?

- How do the assessment instruments support student demonstration of depth of mathematical understanding of the essential learning standards?

Teams should ask these questions throughout the reflective process as members consider the unit tasks and assessment items, and they should also use the questions when planning the next unit. Student assessment performance should be a consideration for revisions in both instructional and assessment practices for the same unit next year, as well as for any adjustments in the following unit.

Directions: Review your short-term SMART goals for the unit's essential learning standards.

1. How should your team respond to the success demonstrated in this scenario?

2. In what ways does your team define its expectation of student success prior to implementing the next unit?

3. How will your team respond to the students who did not achieve proficiency in this unit?

4. What SMART goal has your team set for the essential learning standards in the next unit? Why?

Figure 3.9: Collaborative team SMART goal-setting reflections.

Visit **go.solution-tree.com/mathematicsatwork** to download a reproducible version of this figure.

Team Progress

It is helpful to diagnose your collaborative teams' current reality and action after launching the unit. Ask each team to individually assess its progress on the tenth high-leverage team action using the status check tool in table 3.2. Discuss your perception of your teams' progress on ensuring evidence-based adult goal setting and action for the next unit of study. It matters less which stage your teams are at and more that team members are committed to working together and understanding how to support students who still need help after the end-of-unit assessment as teams seek stage IV—sustaining.

Table 3.2: After-the-Unit-Ends Status Check Tool for HLTA 10—Ensuring Evidence-Based Adult Goal Setting and Action for the Next Unit of Study

Directions: Discuss your perception of your team's progress on the tenth high-leverage team action—ensuring evidence-based adult goal setting and action for the next unit of study. Defend your reasoning.			
Stage I: Pre-Initiating	**Stage II: Initiating**	**Stage III: Developing**	**Stage IV: Sustaining**
We do not set team proficiency targets for each essential learning standard.	We consider end-of-unit assessment results, but we do not set proficiency targets for each essential learning standard.	We independently use student end-of-unit assessment results to determine if proficiency targets were achieved.	We collaboratively use student end-of unit assessment results to determine if proficiency targets were achieved.
We do not know the end-of-unit results of other team members.	We provide some constructive feedback to each other at the end of a unit if asked.	We score some assessments together and calibrate for scoring accuracy.	We score all assessments together and calibrate for scoring accuracy.
We do not analyze end-of-unit results.	We analyze end-of-unit results, but this does not influence our planning for the next unit.	We carefully and independently consider how end-of-unit results impact our planning for the next unit.	We collaboratively and carefully consider how end-of-unit results impact our team planning for the next unit.

Visit **go.solution-tree.com/mathematicsatwork** to download a reproducible version of this table.

Setting Your After-the-Unit Priorities for Team Action

As part of a PLC culture, your collaborative teamwork is never over, even after the unit ends. You and your teams have worked hard to pursue the two high-leverage team actions we outlined in this chapter.

> HLTA 9. Ensuring evidence-based student goal setting and action for the next unit of study
>
> HLTA 10. Ensuring evidence-based adult goal setting and action for the next unit of study

As teams, your teachers should reflect together on the stage they identified for each of these team actions. Based on the results, what should be each team's priority? You can use figure 3.10 to focus teacher time and energy on actions that are most urgent as the unit ends. Focus on these few but complex tasks, and make them matter at a deep level of implementation.

Directions: Identify the stage you rated your team for each of the two high-leverage team actions, and provide a brief rationale.

9. Ensuring evidence-based student goal setting and action for the next unit of study

   Stage I: Pre-Initiating   Stage II: Initiating   Stage III: Developing   Stage IV: Sustaining

   Reason: _____

10. Ensuring evidence-based adult goal setting and action for the next unit of study

   Stage I: Pre-Initiating   Stage II: Initiating   Stage III: Developing   Stage IV: Sustaining

   Reason: _____

With your collaborative team, respond to the red light, yellow light, and green light prompts for the high-leverage team actions that you and your team believe are most urgent to focus on.

Red light: Indicate one activity you will stop doing that limits effective implementation of the high-leverage team actions.

Yellow light: Indicate one activity you will continue to do to be effective with the high-leverage team actions.

Green light: Indicate one activity you will begin to do immediately to become more effective with the high-leverage team actions.

Figure 3.10: Setting your collaborative team's after-the-unit priorities.

Visit **go.solution-tree.com/mathematicsatwork** to download a reproducible version of this figure.

Teacher team after-the-unit work consists of ensuring three items.

> 1. Each student receives effective feedback and uses the common assessment as a formative learning opportunity.
>
> 2. Consistent, valid, and accurate scoring of the common assessment instrument
>
> 3. Teams use the results of the common assessment for adult goal setting and action for the next unit of study and beyond.

These pursuits address steps four and five of the PLC teaching-assessing-learning cycle (see figure 3.1, page 102) and will help teachers prepare for the rigors and the challenge of teaching and learning in the next unit.

The effective implementation of step five (your collaborative teams using ongoing assessment feedback to improve instruction) leads you and your teams to cycle back to step one and begin the process again with the next unit, setting new short-term goals and shifting focus to the next unit of instruction. The cycle is a continuous-improvement process necessary for effective mathematics teaching, assessing, and learning.

What is wonderful about the process of teaching and learning is that both students and teachers improve their understanding of mathematical content, Mathematical Practices and processes, and their role in the formative assessment process.

This chapter completes your investigation of the PLC teaching-assessing-learning cycle and the ten high-leverage team actions that deliver on the promise of improved student achievement.

EPILOGUE

Taking Your Next Steps

So now what? Your collaborative teams have moved through the five stages of the PLC teaching-assessing-learning cycle and should now be ready to start the process again with the next unit. Some of the considerations from this handbook relative to your teams' work with the instructional unit include:

- Was the size of the unit manageable within the teaching-assessing-learning cycle?
- How did the team discussion of essential learning standards help support student understanding?
- How did the design of the mathematical tasks and assessment instruments work? Were they aligned? Did they require demonstrations of student understanding?
- How did the unit formative assessment plan fit with the end-of-unit assessment?
- What was the student and teacher response at the end of the unit?
- Did the team have the proper amount of time needed to complete its work?

The daily expectations of preparing class, scoring and grading student assessments, dealing with students who need extra help, meeting with parents, and having other school meetings often overwhelm teachers. Part of your responsibility as a leader within a PLC culture is to provide the team time necessary to support their work around the ten high-leverage team actions.

Figure E.1 (page 128) provides a final summative evaluation you can use with each team at the beginning and the end of the school year. Use it to identify your teams' current progress on each of the high-leverage team actions and to target specific areas for improvement and focus as you celebrate teams' strengths and help them to prioritize their areas for continued growth.

Reflect on where your teams fall along the continuum for each high-leverage team action. The process of collaboration capitalizes on the fact that you come together to use diverse experiences and knowledge to create a whole that is larger than the sum of the parts. Your effective collaboration around these actions is *the* solution to sustained professional learning—an ongoing and never-ending process of teacher growth, as outlined in the professionalism principle in *Principles to Actions* (NCTM, 2014), necessary to meet Common Core–type mathematics expectations and beyond.

The National Board for Professional Teaching Standards' (2010) mathematics standards state it like this:

> Seeing themselves as partners with other teachers, they are dedicated to improving the profession. They care about the quality of teaching in their schools, and, to this end, their collaboration with colleagues is continuous and explicit. They recognize that collaborating in a professional learning community contributes to their own professional growth, as well as to the growth of their peers, for the benefit of student learning. Teachers promote the ideal that working collaboratively increases knowledge, reflection, and quality of practice and benefits the instructional program. (p. 75)

The new paradigm for the professional development of your mathematics teachers requires an understanding that the knowledge capacity of every teacher matters. More importantly, however, is that every teacher *acts* on that knowledge and transfers the professional development that he or she receives into his or her daily classroom practice.

Assessing Your High-Leverage Team Actions

Directions: Rate your team on a scale of 1 (low) to 6 (high) for your current implementation of each of the ten high-leverage team actions.

Before the Unit (Step 1 of the Cycle)

HLTA 1. We agree on the expectations and intent of the common essential learning standards and Mathematical Practices for the unit.

Rating: _____

Reason: _____

_____

HLTA 2. We identify and discuss student use of higher-level-cognitive-demand mathematical tasks as part of the instruction during the unit.

Rating: _____

Reason: _____

_____

HLTA 3. We develop high-quality common assessment instruments for the unit.

Rating: _____

Reason: _____

_____

HLTA 4. We develop accurate scoring rubrics and proficiency expectations for the common assessment instruments.

Rating: _____

Reason: _____

_____

HLTA 5. We plan and use common homework assignments.

Rating: _____

Reason: _____

_____

During the Unit (Steps 2 and 3 of the Cycle)

HLTA 6. We develop student proficiency in each Mathematical Practice through in-class, higher-level-cognitive-demand mathematical tasks.

Rating: _____

Reason: _____

_____

HLTA 7. We use in-class formative assessment processes effectively.

Rating: _____

Reason: _____

_____

HLTA 8. We use a lesson-design process for lesson planning and collective team inquiry.

Rating: _____

Reason: _____

_____

continued →

After the Unit (Steps 4 and 5 of the Cycle)

HLTA 9. We ensure evidence-based student goal setting and action for the next unit of study.
Rating: _____

Reason: _____

_____

HLTA 10. We ensure evidence-based adult goal setting and action for the next unit of study.
Rating: _____

Reason: _____

_____

Setting Your Collaborative Team Monitoring Priorities

Directions: Review your ratings for the ten high-leverage team actions under the three categories of essential PLC team commitments.

For each category, list your top two or three specific areas for improvement. Based on your knowledge of the team, what should be your focus for adult growth and improvement and knowledge capacity building? Be specific and use your ratings to inform your choices. Also examine your current progress on team-level SMART goals.

1. PLC Teacher Team Agreements for Teaching and Learning Before the Unit Begins

2. PLC Teacher Team Agreements for Teaching and Learning During the Unit

3. PLC Teacher Team Agreements for Teaching and Learning After the Unit Ends

4. PLC Teacher Team Agreements for SMART Goals

What data targets beckon for improved student achievement in each course or grade level for next year? Consider local (unit by unit), state (CCSS mathematics or otherwise), and national data-improvement targets.

Figure E.1: Tool for assessing your actions and setting team priorities.

Visit **go.solution-tree.com/mathematicsatwork** to download a reproducible version of this figure.

APPENDIX A

Standards for Mathematical Practice

Source: NGA & CCSSO, 2010, pp. 6–8. © 2010. National Governors Association Center for Best Practices and Council of Chief State School Officers. All rights reserved. Used with permission.

The Standards for Mathematical Practice describe varieties of expertise that mathematics educators at all levels should seek to develop in their students. These practices rest on important "processes and proficiencies" with longstanding importance in mathematics education. The first of these are the NCTM process standards of problem solving, reasoning and proof, communication, representation, and connections. The second are the strands of mathematical proficiency specified in the National Research Council's report *Adding It Up:* adaptive reasoning, strategic competence, conceptual understanding (comprehension of mathematical concepts, operations and relations), procedural fluency (skill in carrying out procedures flexibly, accurately, efficiently and appropriately), and productive disposition (habitual inclination to see mathematics as sensible, useful, and worthwhile, coupled with a belief in diligence and one's own efficacy).

1. **Make sense of problems and persevere in solving them.** Mathematically proficient students start by explaining to themselves the meaning of a problem and looking for entry points to its solution. They analyze givens, constraints, relationships, and goals. They make conjectures about the form and meaning of the solution and plan a solution pathway rather than simply jumping into a solution attempt. They consider analogous problems, and try special cases and simpler forms of the original problem in order to gain insight into its solution. They monitor and evaluate their progress and change course if necessary. Older students might, depending on the context of the problem, transform algebraic expressions or change the viewing window on their graphing calculator to get the information they need. Mathematically proficient students can explain correspondences between equations, verbal descriptions, tables, and graphs or draw diagrams of important features and relationships, graph data, and search for regularity or trends. Younger students might rely on using concrete objects or pictures to help conceptualize and solve a problem. Mathematically proficient students check their answers to problems using a different method, and they continually ask themselves, "Does this make sense?" They can understand the approaches of others to solving complex problems and identify correspondences between different approaches.

2. **Reason abstractly and quantitatively.** Mathematically proficient students make sense of quantities and their relationships in problem situations. They bring two complementary abilities to bear on problems involving quantitative relationships: the ability to decontextualize—to abstract a given situation and represent it symbolically and manipulate the representing symbols as if they have a life of their own, without necessarily attending to their referents—and the ability to contextualize, to pause as needed during the manipulation process in order to probe into the referents for the symbols involved. Quantitative reasoning entails habits of creating a coherent representation of the problem at hand; considering the units involved; attending to the meaning of quantities, not just how to compute them; and knowing and flexibly using different properties of operations and objects.

3. Construct viable arguments and critique the reasoning of others. Mathematically proficient students understand and use stated assumptions, definitions, and previously established results in constructing arguments. They make conjectures and build a logical progression of statements to explore the truth of their conjectures. They are able to analyze situations by breaking them into cases, and can recognize and use counterexamples. They justify their conclusions, communicate them to others, and respond to the arguments of others. They reason inductively about data, making plausible arguments that take into account the context from which the data arose. Mathematically proficient students are also able to compare the effectiveness of two plausible arguments, distinguish correct logic or reasoning from that which is flawed, and—if there is a flaw in an argument—explain what it is. Elementary students can construct arguments using concrete referents such as objects, drawings, diagrams, and actions. Such arguments can make sense and be correct, even though they are not generalized or made formal until later grades. Later, students learn to determine domains to which an argument applies. Students at all grades can listen or read the arguments of others, decide whether they make sense, and ask useful questions to clarify or improve the arguments.

4. Model with mathematics. Mathematically proficient students can apply the mathematics they know to solve problems arising in everyday life, society, and the workplace. In early grades, this might be as simple as writing an addition equation to describe a situation. In middle grades, a student might apply proportional reasoning to plan a school event or analyze a problem in the community. By high school, a student might use geometry to solve a design problem or use a function to describe how one quantity of interest depends on another. Mathematically proficient students who can apply what they know are comfortable making assumptions and approximations to simplify a complicated situation, realizing that these may need revision later. They are able to identify important quantities in a practical situation and map their relationships using such tools as diagrams, two-way tables, graphs, flowcharts and formulas. They can analyze those relationships mathematically to draw conclusions. They routinely interpret their mathematical results in the context of the situation and reflect on whether the results make sense, possibly improving the model if it has not served its purpose.

5. Use appropriate tools strategically. Mathematically proficient students consider the available tools when solving a mathematical problem. These tools might include pencil and paper, concrete models, a ruler, a protractor, a calculator, a spreadsheet, a computer algebra system, a statistical package, or dynamic geometry software. Proficient students are sufficiently familiar with tools appropriate for their grade or course to make sound decisions about when each of these tools might be helpful, recognizing both the insight to be gained and their limitations. For example, mathematically proficient high school students analyze graphs of functions and solutions generated using a graphing calculator. They detect possible errors by strategically using estimation and other mathematical knowledge. When making mathematical models, they know that technology can enable them to visualize the results of varying assumptions, explore consequences, and compare predictions with data. Mathematically proficient students at various grade levels are able to identify relevant external mathematical resources, such as digital content located on a website, and use them to pose or solve problems. They are able to use technological tools to explore and deepen their understanding of concepts.

6. Attend to precision. Mathematically proficient students try to communicate precisely to others. They try to use clear definitions in discussion with others and in their own reasoning. They state the meaning of the symbols they choose, including using the equal sign consistently and appropriately. They are careful about specifying units of measure, and labeling axes to clarify the correspondence with quantities in a problem. They calculate accurately and efficiently, and express numerical answers with a degree of precision appropriate for the problem context. In the elementary grades, students give carefully formulated explanations to each other. By the time they reach high school they have learned to examine claims and make explicit use of definitions.

7. **Look for and make use of structure.** Mathematically proficient students look closely to discern a pattern or structure. Young students, for example, might notice that three and seven more is the same amount as seven and three more, or they may sort a collection of shapes according to how many sides the shapes have. Later, students will see 7×8 equals the well remembered $7 \times 5 + 7 \times 3$, in preparation for learning about the distributive property. In the expression $x^2 + 9x + 14$, older students can see the 14 as 2×7 and the 9 as $2 + 7$. They recognize the significance of an existing line in a geometric figure and can use the strategy of drawing an auxiliary line for solving problems. They also can step back for an overview and shift perspective. They can see complicated things, such as some algebraic expressions, as single objects or as being composed of several objects. For example, they can see $5 - 3(x - y)^2$ as 5 minus a positive number times a square and use that to realize that its value cannot be more than 5 for any real numbers x and y.

8. **Look for and express regularity in repeated reasoning.** Mathematically proficient students notice if calculations are repeated, and look both for general methods and for shortcuts. Upper elementary students might notice when dividing 25 by 11 that they are repeating the same calculations over and over again, and conclude they have a repeating decimal. By paying attention to the calculation of slope as they repeatedly check whether points are on the line through $(1, 2)$ with slope 3, middle school students might abstract the equation $(y - 2)/(x - 1) = 3$. Noticing the regularity in the way terms cancel when expanding $(x - 1)(x + 1)$, $(x - 1)(x^2 + x + 1)$, and $(x - 1)(x^3 + x^2 + x + 1)$ might lead them to the general formula for the sum of a geometric series. As they work to solve a problem, mathematically proficient students maintain oversight of the process, while attending to the details. They continually evaluate the reasonableness of their intermediate results.

Connecting the Standards for Mathematical Practice to the Standards for Mathematical Content

The Standards for Mathematical Practice describe ways in which developing student practitioners of the discipline of mathematics increasingly ought to engage with the subject matter as they grow in mathematical maturity and expertise throughout the elementary, middle and high school years. Designers of curricula, assessments, and professional development should all attend to the need to connect the mathematical practices to mathematical content in mathematics instruction.

The Standards for Mathematical Content are a balanced combination of procedure and understanding. Expectations that begin with the word "understand" are often especially good opportunities to connect the practices to the content. Students who lack understanding of a topic may rely on procedures too heavily. Without a flexible base from which to work, they may be less likely to consider analogous problems, represent problems coherently, justify conclusions, apply the mathematics to practical situations, use technology mindfully to work with the mathematics, explain the mathematics accurately to other students, step back for an overview, or deviate from a known procedure to find a shortcut. In short, a lack of understanding effectively prevents a student from engaging in the mathematical practices.

In this respect, those content standards which set an expectation of understanding are potential "points of intersection" between the Standards for Mathematical Content and the Standards for Mathematical Practice. These points of intersection are intended to be weighted toward central and generative concepts in the school mathematics curriculum that most merit the time, resources, innovative energies, and focus necessary to qualitatively improve the curriculum, instruction, assessment, professional development, and student achievement in mathematics.

Standards for Mathematical Practice Evidence Tool

Source: © 2013 by Mona Toncheff & Timothy D. Kanold. All rights reserved.
Source for Mathematical Practices: NGA & CCSSO, 2010, pp. 6–8.

Mathematical Practice 1: "Make Sense of Problems and Persevere in Solving Them"

Students:

- Check intermediate answers, and change strategy if necessary
- Think about approaches to solving the problem before beginning
- Draw pictures or diagrams to represent given information
- Have the patience to complete multiple examples in trying to identity a solution
- Start by working a simpler problem
- Make a plan for solving the problem

In the classroom:

- Student teams or groups look at a variety of solution approaches and discuss their merits.
- Student teams or groups compare two different approaches to look for connections.
- Students discuss with their peers whether a particular answer is possible in a given situation and explain their thinking.
- Time is allotted for individual thinking and for sharing thoughts and ideas with a student peer.

Mathematical Practice 2: "Reason Abstractly and Quantitatively"

Students:

- Connect numbers and symbols with a concrete model
- Write and manipulate symbols to solve a problem
- Generalize a pattern (such as, "I see that $i^2 = -1$, $i^3 = -i$, $i^4 = 1$, so i^{257} must be _____ since the pattern is _____")
- Approximate the answer by reasoning quantitatively before doing the actual calculation

In the classroom:

- Students explain the meaning of a given expression or equation in terms of a situation to their peers.
- Students interpret an answer in the context of a problem and verify their reasoning with a peer.

- Students discuss with peers and choose the most likely answer to a given situation without actually doing any calculations.

Mathematical Practice 3: "Construct Viable Arguments and Critique the Reasoning of Others"

Students:

- Explain what is wrong in a process that has led to an incorrect answer
- Justify why an answer makes sense or is reasonable
- Justify each step of a proof argument
- Explain why they chose one possible answer over another for a given problem
- Use a specific example to test a general conjecture

In the classroom:

- Students explain their thinking process to those in their team or group.
- Students justify why they agree or disagree with an answer.
- Students discuss different ways to represent a pattern that they see and explain why their representation works to their peers.
- Students group objects that are similar and explain why they think so.
- Students debate with their peers whether arguments in a proof are mathematically reasonable.

Mathematical Practice 4: "Model With Mathematics"

Students:

- Model a real-world scenario with a mathematical representation
- Use a variety of modalities to represent different scenarios
- Represent a story with concrete objects
- Draw a picture to illustrate a problem
- Analyze data to determine whether a conclusion is reasonable

In the classroom:

- Students work in pairs or groups to design a mathematical model for a given situation.
- Student teams and groups act out a problem scenario or use manipulatives to demonstrate it.
- Students brainstorm the mathematics they have previously learned that they need to represent and solve the situation.

Mathematical Practice 5: "Use Appropriate Tools Strategically"

Students:

- Use algebra tiles to investigate how a pattern develops

- Use compass rulers, protractors, and patty paper to investigate the properties of a geometric figure

- Use graphs and other visual representations to help determine rules about expressions and equations

- Apply relevant formulas correctly to a situation (for example, area, volume, quadratic, and distance)

- Use geometric software to test conjectures

- Use graphing calculators for tasks that are inappropriate to do by hand (like graphing complicated functions, calculating intersections of two graphs, finding a least squares regression line, and so on)

In the classroom:

- Teachers demonstrate appropriate and inappropriate uses of a variety of tools.

- Tools are available for students to use when needed.

- Students share how they chose and used various tools in their solutions.

Mathematical Practice 6: "Attend to Precision"

Students:

- Specify units of measure in their answers

- Define symbols and variables that they are using

- Use vocabulary appropriately

- Measure accurately

- Calculate precisely

- Label graphs and tables correctly

- Recognize and discard extraneous solutions

In the classroom:

- Teacher models appropriate use of mathematical vocabulary.

- Teacher models calculating with units throughout a problem, not just adding them on at the end.

- Student teams or groups discuss appropriate accuracy of answers.

- Student teams or groups discuss differences between a sketch and a graph.

- During student team or group presentations, other class members and the teacher help to clarify language and explanations.

Mathematical Practice 7: "Look for and Make Use of Structure"

Students:

- Look for efficient solution strategies

- Recognize common properties of geometric figures and use them to find answers
- Recognize how to use properties of numbers or equations to simplify a solution
- Identify the sign or the magnitude that a particular answer must have by recognizing a structure in the calculation (for example, answer must be negative or answer must be less than 1)
- Draw auxiliary lines on a geometric figure to reveal a characteristic
- Perform mental arithmetic by rearranging or separating terms to make the calculating easier (for example, in 97 + 105 = 100 + 102, moving 3 to the 97 to make 100)

In the classroom:

- Teacher demonstrates the structure in a complicated expression to help in simplifying it.
- Student teams or groups explain shortcut methods.
- Students outline the structure they've identified or recognized for the rest of the group.

Mathematical Practice 8: "Look for and Express Regularity in Repeated Reasoning"

Students:

- Use their experience in previous work to explain why an answer cannot be possible (such as, "The answer can't be more than 1 because you're dividing a smaller number by a larger one")
- Look for a way to express a shortcut to a solution
- Employ the repeated reasoning to skip steps in a solution

In the classroom:

- Teacher listens for aha moments and extends that thinking for the entire class.
- Student teams or groups discuss their observations, looking for common threads and general rules or properties.

Cognitive-Demand-Level Task-Analysis Guide

Source: Smith & Stein, 1998. © 1998, National Council of Teachers of Mathematics. Used with permission.

Table C.1: Cognitive-Demand Levels of Mathematical Tasks

Lower-Level Cognitive Demand	Higher-Level Cognitive Demand
Memorization Tasks • These tasks involve reproducing previously learned facts, rules, formulae, or definitions to memory. • They cannot be solved using procedures because a procedure does not exist or because the time frame in which the task is being completed is too short to use the procedure. • They are not ambiguous; such tasks involve exact reproduction of previously seen material and what is to be reproduced is clearly and directly stated. • They have no connection to the concepts or meaning that underlie the facts, rules, formulae, or definitions being learned or reproduced.	**Procedures With Connections Tasks** • These procedures focus students' attention on the use of procedures for the purpose of developing deeper levels of understanding of mathematical concepts and ideas. • They suggest pathways to follow (explicitly or implicitly) that are broad general procedures that have close connections to underlying conceptual ideas as opposed to narrow algorithms that are opaque with respect to underlying concepts. • They usually are represented in multiple ways (for example, visual diagrams, manipulatives, symbols, or problem situations). They require some degree of cognitive effort. Although general procedures may be followed, they cannot be followed mindlessly. Students need to engage with the conceptual ideas that underlie the procedures in order to successfully complete the task and develop understanding.
Procedures Without Connections Tasks • These procedures are algorithmic. Use of the procedure is either specifically called for, or its use is evident based on prior instruction, experience, or placement of the task. • They require limited cognitive demand for successful completion. There is little ambiguity about what needs to be done and how to do it. • They have no connection to the concepts or meaning that underlie the procedure being used. • They are focused on producing correct answers rather than developing mathematical understanding. • They require no explanations or have explanations that focus solely on describing the procedure used.	**Doing Mathematics Tasks** • Doing mathematics tasks requires complex and no algorithmic thinking (for example, the task, instructions, or examples do not explicitly suggest a predictable, well-rehearsed approach or pathway). • It requires students to explore and understand the nature of mathematical concepts, processes, or relationships. • It demands self-monitoring or self-regulation of one's own cognitive processes. • It requires students to access relevant knowledge and experiences and make appropriate use of them in working through the task. • It requires students to analyze the task and actively examine task constraints that may limit possible solution strategies and solutions. • It requires considerable cognitive effort and may involve some level of anxiety for the student due to the unpredictable nature of the required solution process.

Sources for Higher-Level-Cognitive-Demand Tasks

Common Core Conversation

www.commoncoreconversation.com/math-resources.html

Common Core Conversation is a collection of more than fifty free website resources for the Common Core State Standards in mathematics and ELA.

EngageNY Mathematics

www.engageny.org/mathematics

The site features curriculum modules from the state of New York that include sample assessment tasks, deep resources, and exemplars for grades preK–12.

Howard County Public School System Secondary Mathematics Common Core

https://secondarymathcommoncore.wikispaces.hcpss.org

This site is a sample wiki for a district K–12 mathematics curriculum.

Illustrative Mathematics

www.illustrativemathematics.org

The main goal of this project is to provide guidance to states, assessment consortia, testing companies, and curriculum developers by illustrating the range and types of mathematical work that students will experience upon implementation of the Common Core State Standards for mathematics.

Inside Mathematics

www.insidemathematics.org/index.php/common-core-standards

This site provides classroom videos and lesson examples to illustrate the Mathematical Practices.

Mathematics Assessment Project

http://map.mathshell.org/materials/index.php

The Mathematics Assessment Project (MAP) aims to bring to life the Common Core State Standards in a way that will help teachers and their students turn their aspirations for achieving them into classroom realities. MAP is a collaboration between the University of California at Berkeley; the Shell Centre team at the University of Nottingham; and the Silicon Valley Mathematics Initiative (MARS).

Mathematics Vision Project

www.mathematicsvisionproject.org

The site features integrated high school curriculum modules that include mathematics performance tasks and video modules connected to Khan Academy.

National Council of Supervisors of Mathematics

www.mathedleadership.org/ccss/itp/index.html

This site features collections of K–12 mathematical tasks for illustrating the Standards for Mathematical Practice. The website includes best-selling books, numerous journal articles, and insights into the teaching and learning of mathematics.

National Council of Teachers of Mathematics Illuminations

http://illuminations.nctm.org

This site provides standards-based resources that improve the teaching and learning of mathematics for all students. The materials illuminate the vision for school mathematics set forth in NCTM's *Principles and Standards for School Mathematics, Curriculum Focal Points for Prekindergarten Through Grade 8 Mathematics*, and *Focus in High School Mathematics: Reasoning and Sense Making*.

National Science Digital Library

http://nsdl.org/commcore/math

The National Science Digital Library (NSDL) contains digital learning objects and tasks that are related to specific Common Core State Standards for mathematics.

Partnership for Assessment of Readiness for College and Careers Task Prototypes and New Sample Items for Mathematics

www.parcconline.org/samples/math

This page contains sample web-based practice assessment tasks (released items) for your use.

Smarter Balanced Assessment Consortium Sample Items and Performance Tasks

www.smarterbalanced.org/sample-items-and-performance-tasks

This site contains sample higher-level-cognitive-demand tasks and online test-taking and performance-assessment tasks (released items) for your use in class.

Virginia Department of Education

www.doe.virginia.gov/instruction/mathematics/index.shtml

This site contains mathematical tasks and teacher team materials to use with the tasks for grades 3–12.

Visit **go.solution-tree.com/mathematicsatwork** for continued updates on this resource list.

How the Mathematics at Work High-Leverage Team Actions Support the NCTM *Principles to Actions: Ensuring Mathematical Success for All*

The *Beyond the Common Core: A Handbook for Mathematics in a PLC at Work* series and the Mathematics at Work process include ten high-leverage team actions teachers should pursue collaboratively every day, in every unit, and every year. The goals of these actions are to eliminate inequities, inconsistencies, and lack of coherence so the focus is on teachers' expectations, instructional practices, assessment practices, and responses to student-demonstrated learning. Therefore, the Mathematics at Work process provides support for NCTM's Guiding Practices for School Mathematics as outlined in the 2014 publication *Principles to Actions: Ensuring Mathematical Success for All* (p. 5). Those principles are:

- **Curriculum principle**—An excellent mathematics program includes a curriculum that develops important mathematics along coherent learning progressions and develops connections among areas of mathematical study and between mathematics and the real world.

- **Professionalism principle**—In an excellent mathematics program, educators hold themselves and their colleagues accountable for the mathematical success of every student and for personal and collective professional growth toward effective teaching and learning of mathematics.

- **Teaching and learning principle**—An excellent mathematics program requires effective teaching that engages students in meaningful learning through individual and collaborative experiences that promote their ability to make sense of mathematical ideas and reason mathematically.

- **Assessment principle**—An excellent mathematics program ensures that assessment is an integral part of instruction, provides evidence of proficiency with important mathematics content and practices, includes a variety of strategies and data sources, and informs feedback to students, instructional decisions, and program improvement.

- **Access and equity principle**—An excellent mathematics program requires that all students have access to a high-quality mathematics curriculum, effective teaching and learning, high expectations, and the support and resources needed to maximize their learning potential.

- **Tools and technology principle**—An excellent mathematics program integrates the use of mathematical tools and technology as essential resources to help students learn and make sense of mathematical ideas, reason mathematically, and communicate their ideas.

Table E.1 (pages 144–145) shows how the HLTAs support NCTM's principles.

Table E.1: The HLTAs and NCTM *Principles to Actions*

Mathematics at Work High-Leverage Team Actions	NCTM's Guiding Practices for School Mathematics
HLTA 1. Making sense of the agreed-on essential learning standards (content and practices) and pacing What do we want all students in each grade level or course to know, understand, demonstrate, and be able to do? Procedures are in place that ensure teacher teams align the most effective mathematical tasks and instructional strategies to the content progression established in the overall unit plan components.	**Curriculum principle** **Professionalism principle.** The professionalism principle specifically calls for teachers to collaboratively examine and prioritize the mathematics content and Mathematical Practices that students are to learn. **Teaching and learning principle.** The teaching and learning principle establishes mathematics goals to focus learning. **Tools and technology principle**
HLTA 2. Identifying higher-level-cognitive-demand mathematical tasks Teacher teams choose mathematical tasks that represent a balance of higher- and lower-level cognitive demand for the essential learning standards of the unit of study.	**Teaching and learning principle.** Effective teaching and learning practices include implementing tasks that promote reasoning and problem solving and supporting productive struggle in learning mathematics. **Tools and technology principle**
HLTA 3. Developing common assessment instruments Develop, design, and create common end-of-unit assessments as a team before the unit begins based on high-quality design and test-evaluation tools. Ensure the assessment instruments are aligned with the instructional discussions and practices used during the unit and connected to all aspects of the essential learning standards for the unit.	**Assessment principle** **Professionalism principle.** The professionalism principle specifically calls for teachers to collaboratively develop and use common assessments. **Tools and technology principle** **Access and equity principle**
HLTA 4. Developing scoring rubrics and proficiency expectations for the common assessment instruments Design common scoring rubrics and assessment practices to align with expected student reasoning and proficiency for every essential learning standard of the unit.	**Assessment principle** **Access and equity principle**
HLTA 5. Planning and using common homework assignments Homework should be viewed as a daily opportunity for formative self-assessment and independent practice for students. Homework protocols include limiting the number of daily tasks, providing spaced practice, balancing cognitive-demand levels, providing all assignments to the students in advance of the unit, and carefully aligning the essential learning standards for the unit.	**Assessment principle** **Access and equity principle**

HLTA 6. Using higher-level-cognitive-demand mathematical tasks effectively Teachers provide targeted and differentiated in-class support as students engage in mathematical processes and peer-to-peer discussions for learning by using higher-level-cognitive-demand tasks in every lesson.	**Teaching and learning principle.** Effective teaching and learning practices include implementing tasks that promote reasoning and problem solving and supporting productive struggle in learning mathematics. **Tools and technology principle**
HLTA 7. Using in-class formative assessment processes effectively Teacher teams do deep planning for small-group discourse and peer-to-peer in-class formative assessment processes via meaningful, specific, and timely teacher feedback with subsequent student action. This requires much more than the diagnostic tool of checking for understanding. To be formative, students must receive feedback during class and take action on that feedback. Teachers intentionally use differentiated and targeted scaffolding and advancing Tier 1 RTI supports as students engage in higher-level-cognitive-demand tasks.	**Assessment principle** **Teaching and learning principle.** Effective teaching and learning practices include eliciting and using evidence of student thinking. **Access and equity principle**
HLTA 8. Using a lesson-design process for lesson planning and collective team inquiry Teachers ensure all lesson elements contain successful opportunities for student demonstration of understanding, with feedback and action on student learning. Teachers actively engage in a teacher team–developed and team-designed lesson, observe teachers teaching the lesson, and debrief the lesson in order to learn from colleagues.	**Professionalism principle.** The professionalism principle specifically calls for teachers to collaboratively discuss, select, and implement common research-informed instructional strategies and plans. **Teaching and learning principle.** All lesson designs should draw from the eight research-informed mathematics teaching practices. **Tools and technology principle**
HLTA 9. Ensuring evidence-based student goal setting and action for the next unit of study Teachers and teacher teams require students to correct their errors and identify the essential learning standards that are strengths and weaknesses based on the results of the end-of-unit assessment. Teachers work with students to complete and carry out a plan for improvement and action based on end-of-unit assessment results and outcomes for proficiency.	**Teaching and learning principle.** Effective teaching and learning practices include eliciting and using evidence of student thinking. **Assessment principle** **Access and equity principle**
HLTA 10. Ensuring evidence-based adult goal setting and action for the next unit of study Teachers and teacher teams score all end-of-unit assessments together and calibrate scoring to ensure accuracy and freedom from bias. Teachers work together after the unit to determine if proficiency targets for students were achieved. Teachers collaboratively and carefully consider how end-of-unit results are used to impact instruction and team planning for the next unit.	**Assessment principle** **Professionalism principle.** The professionalism principle specifically calls for teachers to collaboratively develop action plans that they can implement when students demonstrate that they have or have not attained the standards. **Access and equity principle**

References and Resources

An, S. (2004). *The middle path in math instruction: Solutions for improving math education*. Lanham, MD: Scarecrow Education.

Anderson, J. R., Reder, L. M., & Simon, H. A. (1995). *Applications and misapplications of cognitive psychology to mathematics education*. Unpublished paper, Carnegie Mellon University, Department of Psychology, Pittsburgh, PA. Accessed at http://act-r.psy.cmu.edu/papers/misapplied.html on October 1, 2014.

Black, P., & Wiliam, D. (2001). *Inside the black box: Raising standards through classroom assessment*. London: Assessment Group of the British Educational Research Association.

Boston, M. D., & Smith, M. S. (2009). Transforming secondary mathematics teaching: Increasing the cognitive demands of instructional tasks used in teachers' classrooms. *Journal for Research in Mathematics Education, 40*(2), 119–156.

Burris, C. C., Heubert, J. P., & Levin, H. M. (2006). Accelerating mathematics achievement using heterogeneous grouping. *American Educational Research Journal, 43*(1), 137–154.

Butler, R. (1988). Enhancing and undermining intrinsic motivation: The effects of task-involving and ego-involving evaluation on interest and performance. *British Journal of Educational Psychology, 58*(1), 1–14.

Carpenter, T. P., Franke, M. L., & Levi, L. (2003). *Thinking mathematically: Integrating arithmetic and algebra in elementary school*. Portsmouth, NH: Heinemann.

Chappuis, S., & Stiggins, R. J. (2002). Classroom assessment for learning. *Educational Leadership, 60*(1), 40–43.

Collins, J., & Hansen, M. T. (2011). *Great by choice: Uncertainty, chaos, and luck—Why some thrive despite them all*. New York: HarperCollins.

Common Core State Standards Initiative. (2014). *Key shifts in mathematics*. Accessed at www.corestandards.org/other-resources/key-shifts-in-mathematics on July 15, 2014.

Conzemius, A. E., & O'Neill, J. (2014). *The handbook for SMART school teams: Revitalizing best practices for collaboration* (2nd ed.). Bloomington, IN: Solution Tree Press.

Cooper, H. (2008a). *Effective homework assignments* (Research brief). Reston, VA: National Council of Teachers of Mathematics.

Cooper, H. (2008b). *Homework: What the research says* (Research brief). Reston, VA: National Council of Teachers of Mathematics.

Darling-Hammond, L. (2014). Testing to, and beyond, the Common Core. *Principal*, 8–12. Accessed at www.naesp.org/sites/default/files/Darling-Hammond_JF14.pdf on February 7, 2014.

Davies, A. (2007). Involving students in the classroom assessment process. In D. Reeves (Ed.), *Ahead of the curve: The power of assessment to transform teaching and learning* (pp. 31–57). Bloomington, IN: Solution Tree Press.

Dixon, J. K., & Tobias, J. M. (2013). The "whole" story: Understanding fraction computation. *Mathematics Teaching in the Middle School, 19*(3), 156–163.

DuFour, R., DuFour, R., & Eaker, R. (2008). *Revisiting Professional Learning Communities at Work: New insights for improving schools.* Bloomington, IN: Solution Tree Press.

DuFour, R., DuFour, R., Eaker, R., & Karhanek, G. (2010). *Raising the bar and closing the gap: Whatever it takes.* Bloomington, IN: Solution Tree Press.

DuFour, R., DuFour, R., Eaker, R., & Many, T. (2010). *Learning by doing: A handbook for Professional Learning Communities at Work* (2nd ed.). Bloomington, IN: Solution Tree Press.

Dweck, C. S. (2007). *Mindset: The new psychology of success.* New York: Ballantine Books.

EngageNY. (2013). *Algebra I, module 1.* Accessed at https://content.engageny.org/resource/algebra-i-module-1 on May 10, 2014.

Fernandez, C., & Yoshida, M. (2004). *Lesson study: A Japanese approach to improving mathematics teaching and learning.* Mahwah, NJ: Lawrence Erlbaum Associates.

Fisher, D., Frey, N., & Rothenberg, C. (2010). *Implementing RTI with English learners.* Bloomington, IN: Solution Tree Press.

Foster, D. (2008, November). *The real change agents: Building professional learning community.* Presentation at the California Math Conference. Accessed at www.noycefdn.org/documents/math/real changeagents.pdf on March 14, 2014.

Frayer, D. A., Fredrick, W. C., & Klausmeier, H. J. (1969). *A schema for testing the level of mastery.* Madison: Wisconsin Research and Development Center for Cognitive Learning.

Fullan, M. (2008). *The six secrets of change: What the best leaders do to help their organizations survive and thrive.* San Francisco: Jossey-Bass.

Gersten, R., Taylor, M. J., Keys, T. D., Rolfhus, E., & Newman-Gonchar, R. (2014). *Summary of research on the effectiveness of math professional development approaches.* (REL 2014–010). Washington, DC: National Center for Education Evaluation and Regional Assistance.

Gladwell, M. (2008). *Outliers: The story of success.* New York: Little, Brown.

Hattie, J. A. C. (2009). *Visible learning: A synthesis of over 800 meta-analyses relating to achievement.* New York: Routledge.

Hattie, J. A. C. (2012). *Visible learning for teachers: Maximizing impact on learning.* New York: Routledge.

Hattie, J. A. C., & Yates, G. (2014). *Visible learning and the science of how we learn.* New York: Routledge.

Heflebower, T., Hoegh, J. K., Warrick, P, with Clemens, B., Hoback, M., & McInteer, M. (2014). *A school leader's guide to standards-based grading.* Bloomington, IN: Marzano Research.

Henningsen, M., & Stein, M. K. (1997). Mathematical tasks and student cognition: Classroom-based factors that support and inhibit high-level mathematical thinking and reasoning. *Journal for Research in Mathematics Education, 28*(5), 524–549.

Herman, J., & Linn, R. (2013). *On the road to assessing deeper learning: The status of Smarter Balanced and PARCC assessment consortia* (CRESST Report 823). Los Angeles: University of California, National Center for Research on Evaluation, Standards, and Student Testing.

Hiebert, J. S., & Grouws, D. A. (2007). The effects of classroom mathematics teaching on students' learning. In F. K. Lester Jr. (Ed.), *Second handbook of research on mathematics teaching and learning: A project of the National Council of Teachers of Mathematics* (pp. 371–404). Charlotte, NC: Information Age.

Hiebert, J., & Stigler, J. W. (2000). A proposal for improving classroom teaching: Lessons from the TIMSS video study. *The Elementary School Journal, 101*(1), 3–20.

Jackson, K., Garrison, A., Wilson, J., Gibbons, L., & Shahan, E. (2013). Exploring relationships between setting up complex tasks and opportunities to learn in concluding whole-class discussions in middle-grades mathematics instruction. *Journal for Research in Mathematics Education, 44*(4), 646–682.

Johnson, D. W., & Johnson, R. T. (1999). Making cooperative learning work. *Theory Into Practice, 38*(2), 67–73.

Johnson, D. W., Johnson, R. T., & Holubec, E. J. (2008). *Cooperation in the classroom* (8th ed., rev. ed.). Edina, MN: Interaction Book.

Kagan, S. (1994). *Cooperative learning.* San Clemente, CA: Kagan.

Kagan, S., & Kagan, M. (2009). *Kagan cooperative learning.* San Clemente, CA: Kagan.

Kanold, T. D. (2011). *The five disciplines of PLC leaders.* Bloomington, IN: Solution Tree Press.

Kanold, T. D. (Ed.), Briars, D. J., Asturias, H., Foster, D., & Gale, M. A. (2013). *Common Core mathematics in a PLC at Work, grades 6–8.* Bloomington, IN: Solution Tree Press.

Kanold, T. D., Briars, D. J., & Fennell, F. (2012). *What principals need to know about teaching and learning mathematics.* Bloomington, IN: Solution Tree Press.

Kanold, T. D. (Ed.), Kanold, T. D., & Larson, M. R. (2012). *Common Core mathematics in a PLC at Work, leader's guide.* Bloomington, IN: Solution Tree Press.

Kanold, T. D. (Ed.), Larson, M. R., Fennell, F., Adams, T. L., Dixon, J. K., Kobett, B. M., & Wray, J. A. (2012a). *Common Core mathematics in a PLC at Work, grades K–2.* Bloomington, IN: Solution Tree Press.

Kanold, T. D. (Ed.), Larson, M., Fennell, F., Adams, T. L., Dixon, J. K., Kobett, B. M., & Wray, J. A. (2012b). *Common Core mathematics in a PLC at Work, grades 3–5.* Bloomington, IN: Solution Tree Press.

Kanold, T. D. (Ed.), Zimmermann, G., Carter, J. A., Kanold, T. D., & Toncheff, M. (2012). *Common Core mathematics in a PLC at Work, high school.* Bloomington, IN: Solution Tree Press.

Kennedy, M. M. (2010). Attribution error and the quest for teacher quality. *Educational Researcher, 39*(8), 591–598.

Kidder, T. (1989). *Among schoolchildren.* Franklin Center, PA: Franklin Library.

Kilpatrick, J., Swafford, J., & Findell, B. (Eds.). (2001). *Adding it up: Helping children learn mathematics.* Washington, DC: National Academies Press.

Lappan, G., & Briars, D. (1995). How should mathematics be taught? In I. M. Carl (Ed.), *Prospects for school mathematics: Seventy-five years of progress* (pp. 131–156). Reston, VA: National Council of Teachers of Mathematics.

Larson, M. R. (2011). *Administrator's guide: Interpreting the Common Core State Standards to improve mathematics education.* Reston, VA: National Council of Teachers of Mathematics.

Little, J. W. (2006). *Professional community and professional development in the learning-centered school.* Arlington, VA: National Education Association.

Martin, T. S. (Ed.). (2007). *Mathematics teaching today: Improving practice, improving student learning* (2nd ed.). Reston, VA: National Council of Teachers of Mathematics.

Marzano, R. J. (2006). *Classroom assessment and grading that work.* Alexandria, VA: Association for Supervision and Curriculum Development.

Marzano, R. J. (2007). *The art and science of teaching: A comprehensive framework for effective instruction.* Alexandria, VA: Association for Supervision and Curriculum Development.

Marzano, R. J. (2010). *Formative assessment & standards-based grading.* Bloomington, IN: Marzano Research.

Moller, S., Mickelson, R. A., Stearns, E., Banerjee, N., & Bottia, M. C. (2013). Collective pedagogical teacher culture and mathematics achievement: Differences by race, ethnicity, and socioeconomic status. *Sociology of Education, 86*(2), 174–194.

Morris, A. K., Hiebert, J., & Spitzer, S. M. (2009). Mathematical knowledge for teaching in planning and evaluating instruction: What can preservice teachers learn? *Journal for Research in Mathematics Education, 40*(5), 491–529.

Mueller, C. M., & Dweck, C. S. (1998). Praise for intelligence can undermine children's motivation and performance. *Journal of Personality and Social Psychology, 75*(1), 33–52.

National Board for Professional Teaching Standards. (2010). *National board certification for teachers: Mathematics standards for teachers of students ages 11–18+.* Arlington, VA: Author.

National Council of Teachers of Mathematics. (1991). *Professional standards for teaching mathematics.* Reston, VA: Author.

National Council of Teachers of Mathematics. (2007). *Mathematics teaching today: Improving practice, improving student learning.* Reston, VA: Author.

National Council of Teachers of Mathematics. (2012). *Fuel for thought.* Accessed at www.nctm.org /uploadedFiles/Journals_and_Books/Books/FHSM/RSM-Task/Fuel_for_Thought.pdf on May 10, 2014.

National Council of Teachers of Mathematics. (2014). *Principles to actions: Ensuring mathematical success for all.* Reston, VA: Author.

National Governors Association Center for Best Practices & Council of Chief State School Officers. (2010). *Common Core State Standards for mathematics*. Washington, DC: Authors. Accessed at www.corestandards.org/assets/CCSSI_Math%20Standards.pdf on February 7, 2014.

O'Neill, J., & Conzemius, A. (2006). *The power of SMART goals: Using goals to improve student learning*. Bloomington, IN: Solution Tree Press.

Partnership for Assessment of Readiness for College and Careers. (2013). *Sample mathematics item: Grade 4 "fraction comparison."* Accessed at www.parcconline.org/sites/parcc/files/Grade4 -FractionComparison.pdf on September 23, 2014.

Pashler, H., Rohrer, D., Cepeda, N. J., & Carpenter, S. K. (2007). Enhancing learning and retarding forgetting: Choices and consequences. *Psychonomic Bulletin and Review, 14*(2), 187–193.

Popham, W. J. (2011). *Transformative assessment in action: An inside look at applying the process*. Alexandria, VA: Association for Supervision and Curriculum Development.

Reeves, D. (2011). *Elements of grading: A guide to effective practice*. Bloomington, IN: Solution Tree Press.

Resnick, L. B. (Ed.). (2006). Do the math: Cognitive demand makes a difference. *Research Points: Essential Information for Education Policy, 4*(2), 1–4. Accessed at www.aera.net/Portals/38/docs /Publications/Do%20the%20Math.pdf on January 22, 2014.

Rohrer, D., & Pashler, H. (2007). Increasing retention without increasing study time. *Current Directions in Psychological Science, 16*(4), 183–186. Accessed at www.pashler.com/Articles /RohrerPashler2007CDPS.pdf on March 10, 2014.

Rohrer, D., & Pashler, H. (2010). Recent research on human learning challenges conventional instructional strategies. *Educational Researcher, 39*(5), 406–412.

Silver, E. (2010). Examining what teachers do when they display their best practice: Teaching mathematics for understanding. *Journal of Mathematics Education at Teachers College, 1*(1), 1–6.

Smarter Balanced Assessment Consortium. (2013a). *Grade 7 task sample 43053*. Accessed at http:// sampleitems.smarterbalanced.org/itempreview/sbac/index.htm on March 24, 2014.

Smarter Balanced Assessment Consortium. (2013b). *Smarter Balanced Assessment Consortium: Practice test scoring guide—Grade 4*. Accessed at http://sbac.portal.airast.org/wp-content/uploads/2013/07 /Grade4Math.pdf on March 24, 2014.

Smarter Balanced Assessment Consortium. (2013c). *Smarter Balanced Assessment Consortium: Practice test scoring guide—Grade 11*. Accessed at http://sbac.portal.airast.org/wp-content/uploads/2014/10 /Grade11Math.pdf on March 24, 2014.

Smith, M. S., Bill, V., & Hughes, E. K. (2008). Thinking through a lesson: Successfully implementing high-level tasks. *Mathematics Teaching in the Middle School, 14*(3), 132–138.

Smith, M. S., & Stein, M. K. (1998). Selecting and creating mathematical tasks: From research to practice. *Mathematics Teaching in the Middle School, 3*(5), 348.

Smith, M. S., & Stein, M. K. (2011). *5 practices for orchestrating productive mathematics discussions*. Reston, VA: National Council of Teachers of Mathematics.

Smith, M. S., & Stein, M. K. (2012). Selecting and creating mathematical tasks: From research to practice. In G. Lappan, M. K. Smith, & E. Jones (Eds.), *Rich and engaging mathematical tasks, grades 5–9* (pp. 344–350). Reston, VA: National Council of Teachers of Mathematics.

Stein, M. K., Grover, B. W., & Henningsen, M. (1996). Building student capacity for mathematical thinking and reasoning: An analysis of mathematical tasks used in reform classrooms. *American Educational Research Journal, 33*(2), 455–488.

Stein, M. K., Remillard, J., & Smith, M. S. (2007). How curriculum influences student learning. In F. K. Lester Jr. (Ed.), *Second handbook of research on mathematics teaching and learning: A project of the National Council of Teachers of Mathematics* (pp. 319–370). Charlotte, NC: Information Age.

Stigler, J. W., & Hiebert, J. (1999). *The teaching gap: Best ideas from the world's teachers for improving education in the classroom.* New York: Free Press.

Wallace, W. V. (2013). *Formative assessment: Benefit for all.* Unpublished master's thesis, University of Central Florida, Orlando.

Warshauer, H. K. (2011). *The role of productive struggle in teaching and learning middle school mathematics.* Doctoral dissertation, University of Texas, Austin.

Webb, N. L. (1997). *Criteria for alignment of expectations and assessments in mathematics and science education* (Research Monograph No. 8). Washington, DC: Council of Chief State School Officers.

Webb, N. L. (2002). *Depth-of-knowledge levels for four content areas.* Accessed at www.allentownsd.org /cms/lib01/PA01001524/Centricity/Domain/1502/depth%20of%20knowledge%20guide%20for %20all%20subject%20areas.pdf on February 27, 2014.

Wiliam, D. (2007). Keeping learning on track: Classroom assessment and the regulation of learning. In F. K. Lester Jr. (Ed.), *Second handbook of research on mathematics teaching and learning: A project of the National Council of Teachers of Mathematics* (pp. 1051–1098). Charlotte, NC: Information Age.

Wiliam, D. (2011). *Embedded formative assessment.* Bloomington, IN: Solution Tree Press.

Index

Common Core Mathematics in a PLC at Work™ series
Edited by Timothy D. Kanold
By Thomasenia Lott Adams, Harold Asturias, Diane J. Briars, John A. Carter, Juli K. Dixon, Francis (Skip) Fennell, David Foster, Mardi A. Gale, Timothy D. Kanold, Beth McCord Kobett, Matthew R. Larson, Mona Toncheff, Jonathan A. Wray, and Gwendolyn Zimmermann

These teacher guides illustrate how to sustain successful implementation of the Common Core State Standards for mathematics. Discover what students should learn and how they should learn it at each grade level. Comprehensive and research-affirmed analysis tools and strategies will help you and your collaborative team develop and assess student demonstrations of deep conceptual understanding *and* procedural fluency.

BKF566, BKF568, BKF574, BKF561, BKF559

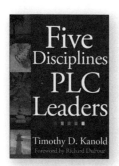

The Five Disciplines of PLC Leaders
By Timothy D. Kanold
Foreword by Richard DuFour

Outstanding leadership in a professional learning community requires practice and patience. Simply trying harder will not yield results; leaders must proactively *train* to get better at the skills that matter. This book offers a framework to focus time, energy, and effort on five key disciplines. Included are reflection exercises to help readers find their own path toward effective PLC leadership.

BKF495

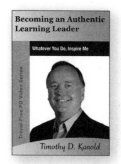

Becoming an Authentic Learning Leader: Whatever You Do, Inspire Me
Featuring Timothy D. Kanold

Encourage your skeptics, cynics, and rebels with Dr. Kanold's eight fundamental disciplines of inspirational leadership. These essential concepts can impact your leadership life as well as the legacy of your leadership team. Practical yet challenging, this humorous and motivational session provides the support and focus needed to sustain effective leadership over time.

DVF063

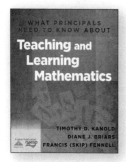

What Principals Need to Know About Teaching and Learning Mathematics
By Timothy D. Kanold, Diane J. Briars, and Francis (Skip) Fennell

Ensure a challenging mathematics experience for every learner, every day. This must-have resource offers support and encouragement for improved mathematics achievement across every grade level of your school. With an emphasis on *Principles and Standards for School Mathematics* and Common Core State Standards, this book covers the importance of mathematics content, learning and instruction, and mathematics assessment.

BKF501

"Tremendous, tremendous, tremendous!

The speaker made me do some very deep internal reflection about the **PLC process** and the personal responsibility I have in making the school improvement process work **for ALL kids.**"

—Marc Rodriguez, teacher effectiveness coach,
Denver Public Schools, Colorado

PD Services

Our experts draw from decades of research and their own experiences to bring you practical strategies for building and sustaining a high-performing PLC. You can choose from a range of customizable services, from a one-day overview to a multiyear process.

Book your PLC PD today!
888.763.9045

Solution Tree

Solution Tree

Solution Tree's mission is to advance the work of our authors. By working with the best researchers and educators worldwide, we strive to be the premier provider of innovative publishing, in-demand events, and inspired professional development designed to transform education to ensure that all students learn.

NATIONAL COUNCIL OF
TEACHERS OF MATHEMATICS

The National Council of Teachers of Mathematics is the public voice of mathematics education, supporting teachers to ensure equitable mathematics learning of the highest quality for all students through vision, leadership, professional development, and research.